HERITAGE RAILWAYS IN THE MIDLANDS

SIMON ELSON

First published 2024

Amberley Publishing
The Hill, Stroud,
Gloucestershire, GL5 4EP

www.amberley-books.com

Copyright © Simon Elson, 2024

The right of Simon Elson to be identified as the Author
of this work has been asserted in accordance with the
Copyright, Designs and Patents Act 1988.

All rights reserved. No part of this book may be reprinted
or reproduced or utilised in any form or by any electronic,
mechanical or other means, now known or hereafter invented,
including photocopying and recording, or in any information
storage or retrieval system, without the permission in writing
from the Publishers.

ISBN: 978 1 3981 1441 8 (print)
ISBN: 978 1 3981 1442 5 (ebook)

British Library Cataloguing in Publication Data.
A catalogue record for this book is available from the British Library.

Typeset in 10pt on 13pt Celeste.
Origination by Amberley Publishing.
Printed in the UK.

Contents

Introduction		5
Chapter 1	The Battlefield Line: Shackerstone to Shenton	7
Chapter 2	Peak Rail: Rowsley to Darley Dale (Derbyshire Peak District)	16
Chapter 3	Rocks by Rail: Cottesmore, Near Oakham	26
Chapter 4	Midland Railway – Butterley: Derbyshire Peak District	34
Chapter 5	Didcot Railway Centre: Didcot, Oxfordshire	42
Chapter 6	Great Central Railway: Loughborough to Leicester	53
Chapter 7	Severn Valley Railway: Kidderminster	61
Chapter 8	Keighley and Worth Valley Railway: West Yorkshire	71
Chapter 9	Foxfield Railway: Near Stoke, Staffordshire	80
Chapter 10	The Train without a Track: Outside Crewe Heritage Centre	88
Chapter 11	A Track without a Train: National Memorial Arboretum, Near Alrewas, Staffordshire	92
Epilogue		96

Introduction

Trains have always fascinated me. During my childhood, local pubs in my hometown of Burton-on-Trent would club together and charter a train for an away day for the regulars. The pubs had a carriage or two each, often BR Mark 1 corridor ones. Crates of Bass, Pedigree, or Double Diamond were loaded on board by your pub depending on which of the town's brewery they were tied to. The trains then made their way to the seaside, as it was the Midlands – either Skegness, Great Yarmouth, or occasionally Blackpool were the usual destinations.

I believe the practice has died out now, mainly due to safety reasons I assume, as a lot of the passengers arrived at the destination extremely drunk and spent most of the day sitting in a deck chair sleeping it off only to repeat the drinking on the way home. I went on a few of these trips as a child (my parents genuinely wanted a day out at the seaside and not an extended drinking session) and I went on what must have been one of the final trips in my late teens with a mate to Blackpool Pleasure Beach – after a few early beers on a train, not a good idea.

Later in our twenties, we discovered a local rail ticket called the Derbyshire Wayfarer, which still exists today. In the nineties it cost around a fiver and allowed unlimited travel for a day on the trains and buses of Derbyshire. Importantly for us, though, tacked on the bottom of the region was Burton-on-Trent and at the top Sheffield. An early train got us to first orders in South Yorkshire and then several other trains took us around Derbyshire, arriving back in our hometown for last orders. Is a pub crawl by rail sensible? Probably not. Great fun? Absolutely.

Having been brought up in the East Midlands, I also got excessive coverage of the APT (Advanced Passenger Train) on the news as it was built mainly in Derby, which was something else that added to my fascination with trains.

'What If' and 'Alternative History' thrillers are all the rage in the book charts as I write this. I've got a 'What If' scenario that's never been done. Brunel's wider gauge becomes the selected width by the 1846 government gauge enquiry and broad gauge becomes the norm for the United Kingdom's rail network. Straighter lines, faster trains. HST speeds two generations ago. Not a plot for a bestselling thriller but an interesting scenario for someone like myself who is obsessed with trains. Spoiler alert: I visit a section of broad gauge and view two broad-gauge trains.

I drive mostly where I want to go now, but if the Beeching axe hadn't fallen in the sixties, would we now be as reliant on cars and motorways as we are? A line near me hasn't run a passenger train in fifty years but is used occasionally by goods trains. If it reopened, would I use it? It links Burton and Leicester, so probably as I support Leicester City FC and hate driving around Leicester. A plan to reopen it has been around for as long as I can remember, but nothing has happened as yet. Another reason I'd like to ride this particular line is that it crosses the River Trent from Burton-on-Trent to Stapenhill on a viaduct. The bridge is well over a century old but at the outbreak of the Second World War, four pillboxes were added to it in case Britain was invaded, two on each side, pointing up and down the river. I've photographed them with a telephoto lens for a magazine article on pillboxes but would like to ride a train on the line to go past them. They are similar to the more famous examples on Putney Bridge in London.

In this book I've been to several standard-gauge heritage railways, mostly running on axed branch lines of railway in the United Kingdom. The heritage railways I visited are not an exhaustive list of Midland railways, others (along with equally fascinating narrow-gauge railways) exist in the Midlands and perhaps I've stretched the Midlands a little to include some railways I particularly wanted to visit. The final two shorter chapters aren't railways as such, but both have stories I wanted to tell and both can be visited, so with a little indulgence on my part I included them.

Read about the history of these lines, my time on board their trains and in the stations and why not visit some of them.

Author's Note

All the heritage railways in this book provided immense help with information and it would be impossible to list all the volunteers and staff who assisted me with research without risking leaving someone out, so a warm thanks to everyone who took the time to meet with me and guide me around the railways.

Some of the trains, rolling stock and railway memorabilia mentioned are not on public display but were made available to me for the purpose of this book. If you read about a particular exhibit that piques your interest, then please check with the railway concerned to make sure the item can be viewed by the public as exhibits are rotated and loaned to other organisations, etc. If it is in their collection but currently not on display, a polite request to go 'behind the scenes' to view it may yield results; it's often just a question of room, for every train, carriage or guards van being used for passengers, there's probably five or six in storage elsewhere with a story just as interesting as the ones in use. Also, where I mention a special event or demonstration day, then again check with the railway for dates.

Chapter 1

The Battlefield Line: Shackerstone to Shenton

Do you remember when Richard III was found in a car park in Leicester? Or perhaps Lord Percy cutting off the king's head played by Peter Cooke in the original *Blackadder* series, with Baldrick muttering the classic line, 'Dear oh dear … Richard the Third'. The reason Richard was in a Leicester car park is that he was killed at the Battle of Market Bosworth a few miles up the road. The Battlefield Line takes its name from that event, running very close to the actual battlefield and visitors' centre.

The story of the line goes back to the canal age with the mining industry developing in the area during the 1700s. A 10-mile-square coal field was established in the area between Leicestershire and South Derbyshire, which was centred around Ashby de la Zouch. The coal was needed around the Midlands and further afield, so a canal was planned in 1796 and completed in 1804, some of the route being covered by horse-drawn tram to cut costs. In 1846 the Midland Railway bought the canal and immediately started to plan a railway line to supplement it. They obtained permission with the condition that they build a branch line to Hinckley. This was not acted upon and had lapsed by 1865, so a new proposal was put forward connecting, at the southern end, Hinkley to the South Leicestershire Railway and, at the northern end, linking Midland Railway's Leicester to Burton-on-Trent line at Moira. It was also suggested that Midland Railway would build a branch line from Shackerstone to Hugglescote, again joining the Burton to Leicester line. A rival proposal by London & North Western Railway (LNWR) for a more comprehensive 50-mile line was also submitted. In 1867 an Act was obtained for a joint construction of the proposed line by both Midland Railway and LNWR. The contract was awarded to Barnes and Beckett for £171,900 (around £20 million in 2022), their bid being the lowest received.

On 18 August 1873, the line opened to goods trains, and passenger services started two weeks later on 1 September. Ten years later the Charnwood Forest line was added linking Coalville to Loughborough, extending the Shackerstone line.

In 1944 the line played an important role in D-Day. A large fuel depot was established at Market Bosworth covering around 100 acres. The depot opened in the summer of 1943 and was initially one of three. The other two were Bottesford, also in Leicestershire, and Barry in Wales. Others were added as D-Day approached; the depot ran a narrow 2-foot gauge.

Fuel was delivered by train, usually consisting of twenty-five tankers. It was decanted into 25-litre jerry cans; millions were manufactured for the forthcoming invasion. The filled jerry cans were distributed by both road and rail to the army units. Many of the tankers still exist and are used by several heritage lines for fuel storage. The depot was closed by 1948 and the site cleared. It is currently agricultural land but is likely to become housing in the near future.

The original line stopped running as a passenger line in 1931, although excursions continued until 1964 and goods trains until 1970. Renovation started on Shackerstone station the same year. In 1973 the line reopened as a heritage line running trains between Shackerstone and Market Bosworth, and in 1992 it was extended to Shenton.

The track is 5 miles long and takes in three stations: Shackerstone, Market Bosworth and Shenton. Of the three stations, two of them are original to the line. Shackerstone station is owned by the railway and was restored in 1970 prior to the line opening. It serves as the main administration centre for the line and houses an impressive railwayana collection, including signs, railway lights, badges, tools, etc .Also on the site are the Victorian Tea Rooms, which have a 1950s feel, Union Jacks (opinion is divided on whether this should be called the Union Jack or Union Flag when flown on land, so I have opted for the popular term) and retro decor making you feel as though you have stepped back in time, possibly to around Queen Elizabeth II's Coronation. When we arrived they seemed to be doing a roaring trade in bacon sandwiches, although we plumped for the rocky road cake.

Across the track on the outbound platform stands the gift shop, stocking the obligatory pencils and pens as well as exquisite railway and other transport models. The shop is housed in a replica of Nuneaton Abbey Street station's waiting room, built from the original plans using similar materials to the original. While it is a recreation, having been built in 1991, it is already over thirty years old in its own right and beginning to look aged.

The signal box at Shackerstone is a relocated building and was originally a canal inspector's office from Measham. The lever assembly usually referred to as a frame was removed from Uttoxeter North Signal Box. The engine shed is in part a relocated cinema from the nearby town of Nuneaton.

Further down the line, Market Bosworth station is not owned or used by the line, but is a car repair garage. A car park is available so the railway can be joined at this point. A waiting room (relocated from Chester Road station, Erdington) and ticket office are also here as well. The stationmaster's house and weighbridge building are still standing but also in private hands and again not used by the railway. The goods shed, however, is rented for use by the railway. A signal box (LNER) at Market Bosworth dating from 1889 was sadly vandalised by arson in 2008. It is currently being restored by the railway.

The end of the line is Shenton. The original station was demolished in the 1940s but a small original lamp room remains and houses a pottery business. The current station was Humberstone Road station from Leicester, which was relocated when the line was extended. The station was rebuilt on the opposite side to the original and flipped around; the original roadside entrance is now trackside and vice versa as it was felt that it fitted the needs of the railway better. Shenton station, as well as a ticket office, houses Kayleigh Young's glass works business, including her showroom, furnace and workshop.

The locomotive in use on our visit was a Class 33, No. 33201, a Bo-Bo layout (four axles in two separate bogies all driven by their own traction motors), and diesel-electric engine.

Built mainly for the Southern Region, it is a more powerful version of the Class 26. As the Southern Region's railways were very seasonal, a larger engine was fitted by removing the steam heater box for the carriages as this would not be required during the popular summer months. In winter their main usage was freight, so no heating was normally required. Ninety-eight units were built by Birmingham Carriage and Wagon Company. Designed in the late 1950s, the first unit, D6500, entered service in January 1960, the last one in 1962.

As well as No. 33201, two Class 20s are in use at present – D8110 and No. 20087. Between 1957 and 1968, 238 Class 20s were built. Class 20s are known as 'Choppers' to rail enthusiasts due to sounding like helicopters when under load.

The railway tends to have several steam engines on loan from various sources, mainly of GWR heritage (Great Western Railway, the original company, not the current heritage railway). On loan to the railway at the time of my visit was CR 419, built in 1907 by Caledonian Railway's St Rollox works in Glasgow. It was in service with British Rail (as No. 55189) until 1962. As one of the last of its type it was saved into heritage railway use by the Scottish Railway Preservation Society (SRPS) in 1964. The purchase price was a not insignificant sum of £750. At some time in its past it had acquired a nonstandard, rather ugly stovepipe chimney. A replacement was found on a sister engine destined for the scrapyard and SRPS restored the engine to its Caledonian Railway's colour scheme and where possible original specification. After various repair work it was removed from service due to its ten-year boiler ticket expiring (a safety certification for the boiler, renewable every ten years). Its final few weeks in service at this time were spent in black BR livery and sporting its BR number (55189) for a photo charter event (where photographers pay to ride the train and get off at interesting locations to take photos of the train). It re-entered service in 2018 and has been 'guest starring' at various heritage railways since then.

The Battlefield line has its own steam engine, *Sir Gomer* (named after Sir James Gomer Berry – the industrial and newspaper magnate (1883–1968)), which is currently out of commission due to its ten-year boiler ticket having expired. It is undergoing the work to return to the track. Built by Peckett & Sons in 1932 and supplied to Mountain Ash Colliery in Llewellyn, it spent all of its working life there, retiring nearly fifty years later in 1981. It was restored and spent a few years at different heritage lines before being purchased by the Battlefield Line in 2001.

The majority of carriages are British Railways Mark 1 corridor carriage stock, the staple of heritage railways. Entering service in 1951 and built until 1974, they were introduced to standardise the carriages across the country after nationalisation of the big four railway companies (Great Western (GWR), Southern (SR), London Midland Scottish (LMS) and London & North Eastern Railways (LNER)). Some buffet cars and corridor carriages are also used – when I visited a dining experience was in full swing, and whisky-tasting trains and 'Santa Specials' in December are also on offer. We travelled on a later Mark 2 open carriage though, one of two added at the front of the train. Built between 1964 and 1975, they were stronger and less prone to corrosion due to a semi-integral construction where the body is attached to chassis to add strength. Some Mark 2 open carriages are still used on the UK rail network as test carriages and brake vans.

The railway is currently working on a passing loop at Market Bosworth so two trains can use the line simultaneously and passenger numbers can increase.

The original Shackerstone station, which was restored in 1970.

The tearoom dressed in period décor from the 1940s/50s, fit for a coronation party.

Shackerstone's platform 1.

Mark 1 and 2 carriages, making up the train on the day we visited.

The gift shop at Shackerstone, a replica of Nuneaton's Abbey Street waiting room, now over thirty years old in its own right.

No. 33201, a Class 33 locomotive with Bo-Bo layout, running via traction motors. A diesel-electric locomotive.

For every coach, guard van or engine in use on a heritage railway dozens more sit ready for restoration or sale – once anything is sold into preservation it is rarely scrapped.

At Shackerstone a large collection of railwayana is on public display.

Market Bosworth waiting room. The actual station is not owned or used by the railway.

Pedestrian footbridge between platforms 1 and 2, Shackerstone station.

14

No. 33201 leaving Shackerstone station pulling both Mark 1 and 2 carriages.

Chapter 2

Peak Rail: Rowsley to Darley Dale (Derbyshire Peak District)

The Peak District is one of the country's most beautiful areas, rivalling the Lake District for stunning scenery. While I love the Lakes, I am biased towards Derbyshire as it is on my doorstep. It is served by two heritage railways.

Peak Rail operates on a line that originally ran from Ambergate to Rowsley in 1849. The line was jointly leased to the Midland and LNWR (London & North Western Railway) in 1852. When a joint route from Rowsley to Buxton could not be agreed by the two railways, Midland Railway got the go-ahead to build their own in 1860. They later got permission to extend the line further from Blackwell to New Mills.

The line opened to Buxton on 30 May 1863. A new station was built at Rowsley, the original one becoming part of the goods yard. In the construction of the line, several major engineering projects were undertaken, which included the building of seven tunnels and two viaducts. In 1866 a line to Manchester London Road (now Manchester Piccadilly) was completed and the line became a through route to Manchester.

Jumping forward to the 1960s, the Manchester connection became important with electrification work on the Manchester to London Euston route closing that line, so the Rowsley line was used as an alternative, using high-speed (for the time) engines called 'The Blue Pullman'. The service was created, providing a first-class link between the two cities. Running between 1960 and 1966, the service ended when the Euston to Manchester line work was complete.

In 1968 the line north of Matlock was closed, but some of it remained in use. In July 1969 the disused Up track section of line between Cromford and Matlock was used for the film *The Virgin and The Gypsy*, based on the book by D. H. Lawrence. Cromford station was included and a steam train, a Peckett 0-4-0T, along with two coaches, was brought in from Keighley and Worth Valley Railway by road for the filming. The line closed totally soon after.

A notable event in Peak Rail's history as a heritage line was when it hosted the *Flying Scotsman* for nine days in 2000. The rolling stock it pulled was also brought in by steam using a GWR Pannier Tank engine, as weight restrictions on bridges between Ambergate and Matlock prevented the use of a heavier engine. At the time, the *Flying Scotsman* was owned Tony Marchington, a pharmaceutical entrepreneur who was well known in heritage circles. He owned many other notable vehicles, one of which was the traction engine *The Iron Maiden*, famous for being the star of the 1963 film of the same name. Tony Marchington also opened the extension to Rowsley, south of the reopened line, in 1997 with a Black Stanier (later known as Black Fives), No. 45337, hauling the opening special and then staying as guest locomotive for the year.

On my visit to Peak Rail, we travelled in a Mark 1 corridor carriage pulled by a Class 44 – D8 *Penyghent*. The railway is lucky to have an engine that spent its working life on the line. Class 44s were the first of the Peak Class of engines and followed by the Classes 45 (more about the Class 45 later in the book) and 46.

Originally designed to be run with a Co-co bogie arrangement (two six-wheeled bogies with all axles powered by a separate motor), these proved fragile and unreliable, suffering from frame fractures. They were replaced with 1 Co-co 1 arrangement (two eight-wheeled bogies with three axles powered on each, the fourth axle unpowered to reduce axle load). Other problems included the vacuum exhausters and the original batteries for the auxiliary motors becoming exhausted quickly. Built between 1958 and 1960, only ten Class 44 locomotives were built before the more powerful Class 45 replaced it. Most were then used for freight work out of Toton. Two remain preserved, both in Derbyshire, this one at Peak Rail and the other at Midland Rail – Butterley.

Disastrous rebranding and expensive logo redesigns, which are then ridiculed by the public, are associated with the modern age of social media, with pictures and memes going viral forcing quick changes. It did happen in the past, though, but they just took longer to be noticed. In 1948 British Rail introduced their first logo (perhaps trademark was more the term then): a lion above a locomotive drive wheel, which was soon described as 'The Lion on a Unicycle'. Among the locomotives at Peak Rail, I spotted a locomotive sporting this design and couldn't resist a smile at its nickname. It lasted until 1956 when it was replaced by a lion holding a spoked driving wheel.

An interesting item at Peak Rail is the LMS carriage that has been painstakingly restored as a dining car. Being made mainly of wood, not many coaches of this era, the 1920s and 1930s, survived and if they did, they were often in a very sorry state. This particular example has a story to tell about its survival. In 1961 at the height of the Cold War, a number of trains were assembled, painted black and hidden in sidings throughout the country. Their title was 'Emergency Response Train'. Each British Rail region supplied two trains usually comprising four modified coaches, including a mobile office, mess, or canteen carriage, generators, and a storage car for specialist equipment, holding supplies such as cable reels and various items to re-establish communications, utilities, etc. in an emergency.

The carriages were ones that had been withdrawn from service after the introduction of the British Rail Mark 1 carriage as they were in plentiful supply. Spending most of their time for the next twenty years maintained and often covered up from prying eyes with tarpaulins in sidings, they survived largely intact until the late 1970s when the trains were stood down. These structurally sound early twentieth-century carriages were ripe for restoration and then used in the burgeoning heritage railway sector, a few remaining in use on heritage railways. The LMS one at Peak, though, is a particularly striking example and can be booked as a dining experience. It was in use as the dining car during my visit and was certainly an impressive sight.

The engine that pulls the LMS dining car is a Bagnall 0-4-0 ST (two axles with four coupled wheels, all of which are driven, and ST indicates a saddle tank where the water tank sits on the top and sides of the boiler) called *Dunlop No. 6*. It was originally built for the Dunlop Rubber Company. Bagnalls were located in Stafford, relatively close to Fort Dunlop in Birmingham. I was allowed onto the footplate while it was getting ready for the dining experience – the furnace was alight and the boiler was getting up steam.

Peak Rail has a full-size turntable in a fully usable state of repair. When the site was taken over the original concrete pit was intact but had been filled in with rubble. This was cleared out and a replacement sourced from the Mold Valley Line in Wales and fitted. It can take most full-size locomotives, including large steam engines. Mainly working from the vacuum pump of the locomotive, the turntable negates the need for large loops of track to turn an engine around. Other methods of moving an engine on the turntable are using the winding gear and the good old-fashioned manual push.

A signal box has been re-erected after being moved from Bamford Hope Valley Line. It is hoped it will be an interactive exhibit when completed. It is thought to be unique in

Rowsley South platform.

preservation circles as it is operated by cables rather than rods. The frame (levers) was donated by the National Railway Museum in York.

Sitting just by the engine shed is a footbridge. It is believed to be a unique survivor of a double staircase and was originally located at Station Road in Darley Dale. It was removed in the early 1970s and a similar one in Church Lane was cut up for scrap. It connected the Up and Down platforms. Ordered in 1910, it was erected in spring of the following year.

After the station road bridge's removal by Derbyshire County Council it was sent to Midland Railway – Butterley. The plan was to re-erect it there in the late 1970s, but this didn't happen and while it was there the cast-iron supports were stolen, presumably to be sold for scrap. In 2009 it was returned to the Peak Rail site and is now owned by Derwent & Wye Valley Railway Trust. The missing supports were recast by Steelway in the West Midlands and are now securely stored ready for the rebuilding process. Planning permission has been granted and it is hoped that after a full renovation it will return to exactly the same location from where it was removed from over half a century ago. It will certainly be an impressive sight and a story of perseverance.

Mark 1 'Corridor' carriage.

D8 *Penyghent*, a Class 44 at Matlock Riverside, spent most of its working life on the line where it is now preserved.

Peak Rail's LMS dining car. It was preserved through its use as a top-secret 'Emergency Response Train' during the Cold War.

Dunlop No. 6, a Bagnall 040 ST, used to pull the LMS dining car on the line for luncheon specials.

On the footplate of *Dunlop No. 6*.

Signal box at Rowsley South (relocated from Bamford Hope Valley Line), thought to be unique in preservation railways as it is operated by cables rather than the usual rods.

The cables from the signal box.

The working full-size turntable relocated from Mold Valley Line in Wales.

This double-staircase footbridge is a unique survivor. It was removed from Station Road in Darley Dale, and will hopefully be restored and replaced at its original site.

Platform at Rowsley South.

Platform sign at Matlock Riverside.

The much-maligned British Railways logo, nicknamed 'Lion on a unicycle'.

Chapter 3

Rocks by Rail: Cottesmore, Near Oakham

'The fireman tells the driver when to drive the guards train on the signalman's railway.' I first heard this phrase at Rocks by Rail, and it reverses the usual accepted hierarchy. It refers to the fact that the train can't move until the pressure is up, so the fireman is the one who starts the procedure. I learnt this phrase while standing on the footplate of one of their steam trains currently under restoration.

Rocks by Rail does have a working steam engine, an Andrew Barclay 0-4-0 locomotive built in 1927. Carrying its works number of 1931, it was delivered to Foley Park for use at West Midlands Beat Sugar Company in their sidings that now forms part of the Severn Valley Railway. On the day of my visit a diesel was in use due to the 'Driver for a Fiver' day and No. 1931 was tucked away.

Rocks by Rail is slightly different to other heritage railways in that it was originally a quarry railway not a passenger railway, but it was connected to the rail network to distribute the stone. The line is 1.4 miles but only ¾ mile are for passenger use. This originally made up the exchange sidings serving Burley Park, Exton Park quarries and Cottesmore West Pit, which was originally a narrow-gauge tram line worked by horses. It later became steam-powered 3-foot gauge.

In 1979 6 acres of land on the end of the Cottesmore Mineral Line were purchased from British Rail for the small collection of engines that had already been collected by the then named Rutland Railway Museum. A short length of track was laid in 1980 and the railway was born.

The engine often used is *Betty* S0201, a Rolls-Royce Sentinel. Two other Sentinels are used by the railway, *Jean* and *Graham*, the male-named Sentinels, while similar specification, are built heavier to allow for hauling greater loads.

Betty was on duty during my visit. Passenger trains are normally run using a brake van rather a coach owing to the line's short distance. However, the museum often run a 'Driver for a Fiver' allowing visitors to drive a diesel locomotive – no other passengers are on board for obvious safety reasons. I drove Betty, a 325 bhp Rolls-Royce, a total of a mile and a half. The driving is very hands on. Going out, the route is downhill, so the brakes have to be applied gently and frequently to stop the engine running away with itself and crashing into the buffers at the end – obviously not a good idea. You are shown what to do before starting off and advised during your session, but the controls are not normally touched by anyone else but you. As a grown man I still found it both exciting and slightly alarming because as an adult you think you can do most things, but occasionally something comes along that puts you way out of your comfort zone. This was one of the two things in recent years that did this (the other being a few laps of a velodrome – that was very scary).

In the engine shed is a registered war memorial in the form of a railway engine, HL 3865 *Singapore*. Built in 1936 by Hawthorn Leslie & Co., who were predominantly shipwrights, it was supplied new to the Royal Navy Dockyard in Singapore. It was still working there at the fall of the base in 1942 and was captured by the Japanese Army and used by them until the end of the war. It still has damage to its bodywork, sustained in the fighting. This is labelled as bullet holes but could also be shrapnel damage. It returned to the UK in 1953 and worked in Chatham Dockyard. Following the move to diesel engines in the early 1970s by the dockyard, it was put into storage and finally purchased by Rocks by Rail in 1978 and became the railway's first working steam engine.

I did learn a little about steam engines that I didn't know while in the engine shed. If a heritage railway is running a steam engine, it takes around three to four hours to warm up, so volunteers are often on site at 6.30 a.m. for a ten o'clock opening. In the days of steam they never fully cooled down, often leaving a small token fire, so the warm-up time was shorter in the mornings.

Another thing I learnt was about a safety feature called a fusible plug, a plate in the crown of the firebox covering a small opening which was soldered on. If the boiler was running out of water, the heat would rise above the safe operating temperature, which would melt the lead solder and deposit the remainder of the water on the fire, hopefully putting the fire out. If it was ever used, then it wasn't a quick fix to repair, requiring an extensive inspection before the engine could be operated again.

It is possible to walk alongside the track and a few interesting rolling stock items are on show. One is a goods wagon that had a relatively recently painted livery on it – Necropolis Railway. The original Necropolis Railway was a short line that ran from Waterloo using a dedicated station (not the current Waterloo station) to Brookwood Cemetery in Surrey, a distance of 23 miles, carrying corpses and mourners. The goods wagon at Rocks by Rail is a recreation using an old wagon for the television series *Ripper Street* (2012–16).

As well as railway engines, I am always interested in anything mechanical that is either huge or rare, and Rocks by Rail had something in their history that was both, immediately grabbing my attention. The name of the museum and railway indicates that there is more to their heritage than just rail. The Exton Park Quarry, roughly 2 miles away, when operational was home to what was once the largest walking dragline excavator in the world, Sundew, which was named after the Grand National winner of 1957, the year the machine was completed and used.

Weighing in at 1,675 tons with a boom length of 282 feet, standing taller than Nelson's Column, it dominated the skyline for miles around. Its bucket was 20 cubic yards and could excavate and transport its own weight every hour, and the cycle of digging and discharging could be completed in sixty seconds. The original home of Sundew was Exton Park Quarry, part of the quarry network that the track at Rocks by Rail served. Exton Park's original estimated output was 10,000 tons per week, 5,000 from each of the two faces, but such was the size of Sundew, the 10,000-ton target was reached by just the south face.

By the mid-1960s the quarry industry was in decline and in 1973 Exton Park was closed. After Sundew helped tidy up the site, including levelling up and filling in, it was left to await its fate.

A new quarry was opening 13 miles away in Harringworth near Corby and it was obvious it would need a dragline of its own. A new lease of life for Sundew was suggested, but the logistics of getting it there were as big as Sundew itself. Forty times larger than the weight limit of a low loader and many times bigger than any lorry, moving by road was impossible, breaking it down and rebuilding it like a giant Meccano set was estimated at around a two-year time scale and at a cost of quarter of a million pounds, a significant sum in 1973.

The motive power of Sundew was supplied by enormous shoes, not tracks as most people would expect, and allegedly somebody joked at a planning meeting about it walking the 13 miles. The joke was soon turned into a reality, and a short feasibility study went into using hovercraft technology but it was decided that the pressure would still damage the ground, so a year's planning went into the walking operation and it took eight weeks. Starting on 30 May 1974, one of the more interesting problems was its power; Sundew operated on a tethered supply, using a lead 1,000 feet long allowing it to walk around its normal environment, but obviously a lot longer range was required. For technical reasons a generator could not be used, so a series of junction boxes were set up along the route using Exton Park Quarry at one end and Harringworth Power Station at the other.

During the walk John Noakes visited and took the controls for a short time. It was filmed and shown on the BBC's *Blue Peter* children's programme.

Sundew finally stopped working just six years after its amazing journey to Wakerly, lowering its gigantic bucket for the last time onto a pile of limestone placed there to take the weight of the boom on 3 January 1980. Some of the other large draglines were shipped to America, but Sundew was boarded up to prevent vandalism and souvenir hunters gaining access until its fate could be decided. In 1987 the cutting torches of the scrap men started to dismantle the unique machine. The operator's cab was saved, brought back to Rocks by Rail and can be seen by visitors. It is placed on a bank high above the railway, and you can even sit in the chair and look over the track below. Plans are in the early stages of perhaps turning the cab into an interactive exhibit showing footage of similar machines in action and charting the history of Sundew on a screen.

Rocks by Rail have a smaller working dragline crane, which is often used on demonstration days, so that visitors can see what used to happen with the limestone removed from the quarry. A 2-cubic-yard bucket is used, a lot smaller than the 120-cubic-yard bucket that could hold four minis that was on Sundew. A scale model is on show in the museum, completed in 1998 by David Corbidge, which shows the enormous size of Sundew as it towers over the model railway and quarry in the same exhibit. While not a true railway attraction, Sundew's history is part of Rocks by Rail and has to be included in this railway's story. Tie in the 'Driver for a Fiver' day, where everyone gets a chance to drive for a tiny fee, then this railway in England's smallest county – Rutland – is an absolute must visit for any enthusiast.

Rocks by Rail is near Oakham, a small town that I have a soft spot for, visiting it often as my wife has family there. The town is worth a visit for Oakham Castle, which is a sizeable manor house with a display of ornamental horseshoes, large, exaggerated ones for decorative purposes, presented in the castle in the past by visiting dignitaries and oddly hung upside down. Legend has it that the Earl Ferrers supplied nails and shoes for the king's cavalry after the demise of Richard III and had the right to demand a horseshoe from any visiting monarch or peer of the realm. The Romans reckoned that hanging a horseshoe open end up collected all luck, both good and bad. So they set their hanging horseshoes pointing horizontally. Rutland's upside-down horseshoes collect no luck at all, good or bad. And while I won't turn this book into a good-pub-style book, but I must mention a brewery tap house, The Grain Store, near Oakham railway station, which is one of my favourite pubs anywhere. Praise indeed when you consider I reside in Burton-on-Trent, the historic brewing capital of the United Kingdom.

Rolls-Royce Sentinel *Betty*, which can be driven on special event days for (at the time of writing) £5 – Driver for a Fiver.

The cab of *Betty*.

The slightly heavier 'male' Sentinel *Graham*.

HL 3865 *Singapore*, a registered war memorial.

Necropolis guards van painted for the TV series *Ripper Street*.

Scale model of Sundew and the area it worked.

Sundew's cab (exterior).

Sundew's cab (interior).

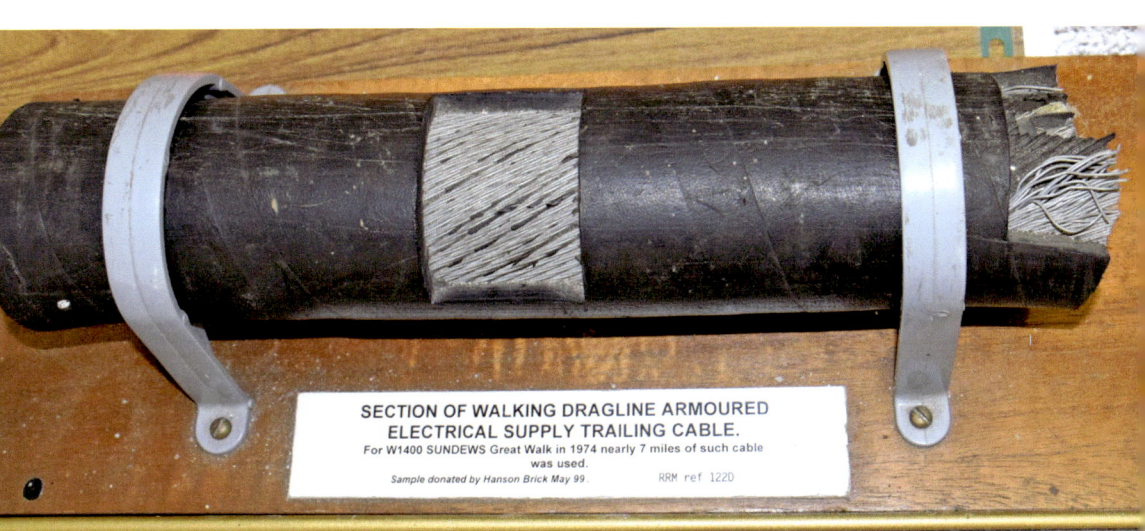

A working dragline used for demonstration days at Rocks by Rail.

The tethered power cable used to supply power to Sundew. A 1,000-foot (330-metre) cable was used in the long walk.

Chapter 4

Midland Railway – Butterley: Derbyshire Peak District

Midland Railway – Butterley operates from Butterley station in the Peak District and runs to Swanwick Junction station. The line's original route was Codnor Park Junction to Crich Junction, and 3½ miles have been relaid for heritage use. The line closed as a British Rail line in 1968. At first Derbyshire County Council was involved and had ideas of creating a leisure activity at previously derelict railway sites. Due to funding issues the council backed out early in the project, leaving volunteers to form the Midland Railway Company in 1973. In 1976 the name changed to Midland Railway Trust.

Work started in 1973, and Butterley was chosen as the base of the railway as it was the site of the only station on the line and ran to Swanwick Junction where the museum would be sited.

The two stations are reclaimed stations. Butterley station was originally sited at Whitwell, Derbyshire, and was rebuilt brick by brick at Butterley in 1981 (Whitwell now has a station and a main-line route serving but is a new station built when the line began to reopen in 1998). Swanwick Junction station was previously Syston station in Leicestershire.

Other buildings include a church, St Saviour's at Swanwick Junction. It was originally at the village of Westhouse. The Midland Railway paid for 50 per cent of its cost in 1898. It became known as the 'Tin Tabernacle' due to its construction type of corrugated iron. It was moved in 1995 to Swanwick Junction and re-erected in just a few months. It is clad inside with wooden planks and can be visited and entered most days the railway is open. Annual General Meetings of the Midland Railway society are held there, and you can have wedding blessings there (although it isn't licensed for actual legal wedding ceremonies). The last time I was at Midland Railway – Butterley for a diesel weekend in the late 1980s, the Tin Tabernacle was still in its original location.

In action when I attended were two relatively new engines for heritage use: a Class 141 and Class 142, one unit at each end. When in use they were usually matched pairs, i.e. two 141s or two 142s. The two different engines do talk to each other when in use together, but a little reluctantly. The Class 141 unit at Midland Railway – Butterley is believed to be the only one saved into preservation. Both are 'Railbus'-type engines sharing bodies and interiors with a production bus. It allowed a new fleet of locos to be designed and built relatively cheaply. British Rail trialled the idea in 1950s with a few entering limited service in the early 1960s. In the 1980s the idea was revived. Known collectively as 'Pacers', the classes 140, 141, 143 and 144 were built. Class 141s locos ran from 1984 to 2005, and Class 142s from 1985 until as recently as 2020. The interiors looked almost identical. They were refurbished after the society was offered nine rolls of moquette, a hard-wearing velour-type material, at a bargain price, each 25 metres long and redundant stock from Bombardier in Derby, giving the society many years of refurbishment and repair to the seats. The material obtained from Bombardier doesn't match exactly the original pattern but as every seat was recovered it didn't matter. Sometimes, though, when originality is paramount, a trust may club together with other groups to get a particular pattern remade.

Midland Railway – Butterley has an exhibition hall that unfortunately is closed to the public at the moment (at time of writing – late 2022) due to urgent repairs being required on the roof. I was, however, given a tour of the building. Some real treasures are in there.

Being an APT (Advanced Passenger Train) fan, something I found particularly interesting was a mock-up cab of a train that didn't make it much past the drawing board. A Class 93 engine, along with five Mark 5 carriages, would have made up the new Intercity 250 planned in the early 1990s for the West Coast Main Line with a design speed of 155 mph. The Class 93 mock-up was made for a trade exhibition in the early 1980s. The Class 93 shown is not to be confused with its more recent namesake currently in production by Stadler in Valencia.

The late Queen Elizabeth II's dining saloon car is also in storage at Midland Railway – Butterley. Built in 1956, it was used extensively during the 1960s and 1970s. It was built at Wolverton Coach Works, which had a long career building royal trains including one for Queen Victoria in 1869. The most recent one they built was in 1977.

The dining car at Midland Railway – Butterley is still laid out with crockery as it would have been while in service. It is based on a Mark 1 carriage but has quarter inch bulletproof glass and the royal train had its own generator car. During the building process particular attention was paid to insulation and sound proofing. A menu is still on display, perhaps from its last journey – smoked salmon/egg mayonnaise as starter, escalope of veal for the main, strawberries and cream for dessert, finishing off with coffee – you would've thought tea.

Sitting in the engine shed at Swanwick is a Class 45 (or more precisely a Class 45/1, the ETH (Electric Train Heat) Connector differentiates the two). This engine is a real flashback to my youth. In my teens these trains were gradually being withdrawn from service and I roamed around the Midlands trying to ride the last few. Walking home from work one Friday, my friend's car screeched to a halt follwed by a shout of 'get in'. I got in to find the rest of the forty-five gang already seated. They had already been to my house to find me, my parents telling them where I was likely to be – this is long before mobile phones became common. I asked the reason for the Bodie-and-Doyle-style drama: 'A forty-five is going through Willington.' One of our group worked on the railways and had seen the schedule earlier in the day while at work. We reached a vantage point just two minutes ahead of the 45 and watched it pass underneath the bridge we were standing on. This would have been the late 1980s and one of the last mainline appearances of a Class 45.

The example at Swanwick is No. 45133, which belongs to The Class 45/1 Preservation Society and is currently undergoing restoration.

No. 45133 was withdrawn from service in May 1987 and was in fully working order apart from faulty batteries and stored at March Traction Maintenance Depot in Cambridgeshire.

The batteries were swapped and it was back in operational condition and kept that way while a sale into preservation was sought. It was purchased in 1990 by the trust and moved to Midland Railway – Butterley. It ran in preservation for the first time on 25 May 1990. It ran for twenty-seven years relatively trouble free. In 2017 it was decided to give the loco a much-deserved overhaul including engine, generator, electrics and a lot of the exterior panels. It is at the time of writing being reassembled and hopefully will have some test runs in the near future and may re-enter service at Midland Railway – Butterley soon. Hopefully, it may visit other preserved railways around the country. I'm planning to visit when its running and ride a Class 45 once again.

Sitting in a siding, I spotted a locomotive that has only recently been retired, an Intercity 125, which formed the backbone of British Rail's high-speed trains over forty years.

The Intercity 125 was built between 1975 and 1982, and ninety-five 'sets' were built (two power cars – Class 43s and Mark 3 carriages). Withdrawal began in 2017. It was originally introduced as a stop-gap until the APT (more about the APT later in the book) but stayed in service for over forty years.

No. 43048 is owned by the 125 group. It was originally delivered in April 1977 to Old Oak Common Depot. It was transferred to Neville Hill Depot in 1982 where it remained for thirty-eight years. It was donated to the 125 group in 2020 and was the first loco to run at 125 mph while in preservation rather than regular service. It is in good working order and can be seen operating at Midland Railway – Butterley. The 125 in service complements the mock-up of the Intercity 250 cab mentioned early.

Swanwick station is also home to Golden Valley Light Railway, a 2-foot narrow gauge railway operating over a mile of track. Running when I visited was a Simplex 60S that was originally delivered to Campbells Brickworks in 1968; it arrived at Golden Valley Light Railway in 1987. It has had a restoration including the fitting of air brakes so it can safely operate a passenger service. It is resplendent having been repainted in its original colour of yellow with Campbells Brickworks markings. Open carriages blew the cobwebs out on the short trip round the surrounding area. A small fare is charged for this.

A museum that claims to be the only one in the world is also at Swanwick station – the National Fork Truck Heritage Centre. Being a transport buff, everything with wheels interests me and holding a fork-truck licence, a museum dedicated to them intrigued me. I was in there showing my family the various controls on the more modern examples that resemble the trucks I've used in the past. I could have done the safety checks on them as well if asked.

The oldest fork truck on display is a Yale E26S/10 dating from 1926. The other exhibits date right up to the 1990s. Entry is free once you have purchased a ticket for the Midland Railway site.

Next door to the National Fork Truck Heritage Centre is the Road Transport Gallery, a small museum which held a gem for me. Most people think of the Routemaster when asked to name a vintage double-decker, but I think of the Bristol Loddeka, a bus that featured in the 1970s sitcom *On the Buses*. The gallery has an example currently undergoing restoration – I was tempted to shout 'I'll get you Butler' and shake my fist at it.

Over 5,000 Lodekkas were made between 1949 and 1969, a longer production run and double the number made than the more famous Routemaster. Is it less famous due to the 'London' effect? Routemasters were mainly used in London and became the 'Red Double

Decker' synonymous with Britain; whereas the Lodekkas were used around the country in different colours, foreign tourists just saw the red Routemasters. And while we are talking about Routemasters, Cliff Richard didn't drive one to Athens in *Summer Holiday*; that was its predecessor an AEC Regent III. All transport is interesting, you just have to find the story.

Butterley station and platform.

Pacer type locomotives. No. 142011 head to head with No. 141113, which is believed to be the only 141 currently in running condition in preservation.

Class 45/1 No. 45133 owned by Class 45/1 Preservation Society.

The 'Tin Tabernacle' at Swanick.

Mock-up of the proposed but never built Class 93.

Queen Elizabeth II's dining car.

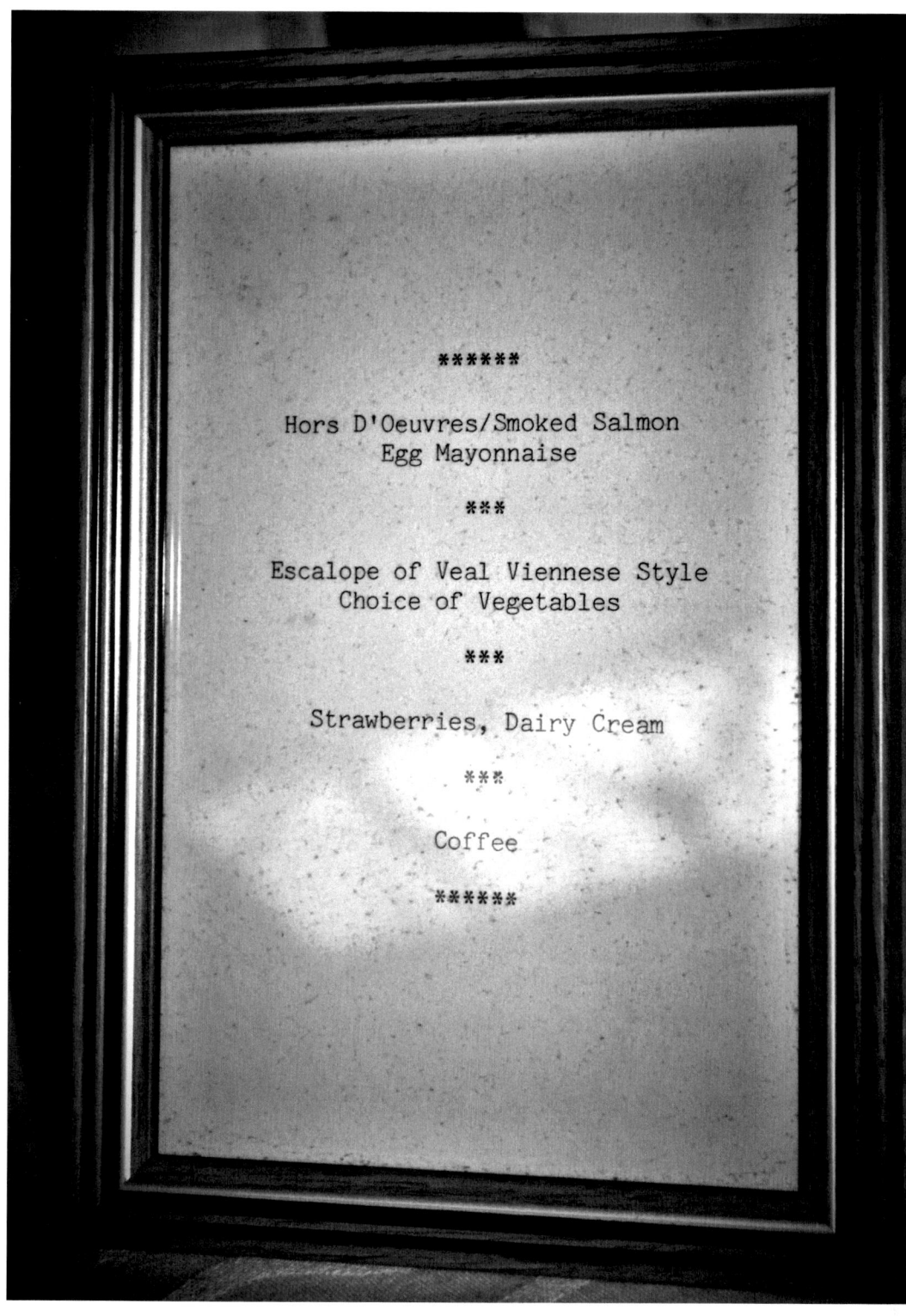

Menu from the dining car.

The National Fork Truck Heritage Centre.

Intercity 125 No. 43048, the first locomotive in preservation to reach 125 mph.

41

Chapter 5

Didcot Railway Centre: Didcot, Oxfordshire

Seven feet and a quarter inch – Isambard Kingdom Brunel's broad gauge. If the battle of the gauges had gone the other way, then the United Kingdom's railway network would be totally different. Wider, more stable, and perhaps faster trains.

The first steam-powered passenger railway opened in 1825 between Stockton and Darlington. Five years later the Liverpool to Manchester line followed it using standard gauge 4 feet 8¼ inches. Legend has it that this gauge was created by George Stephenson after measuring horse-drawn coal trucks already in use in mines and taking an average. In 1833 Brunel, aged just twenty-seven, devised broad gauge for the Great Western Railway (GWR), originally 7 feet, later widened by the quarter inch to perhaps reduce stress round tight bends. By the height of railway mania standard gauge was well established, some routes already terminated in London. Brunel's routes linked the South West, West Midlands and Wales with London. As passengers had to change trains for onward journeys and more importantly goods had to as well, it was decided that one of the gauges had to prevail over the other. It was one of the first format wars predating the VHS/Betamax and CD/Minidisc battles over a century later. Parliament found in favour of standard gauge and broad gauge was gradually fazed out. New railways could still be built if in the geographical region of GWR but elsewhere in the country standard gauge had to be used. The writing was on the wall for Brunel's broad gauge. It continued for a few more decades but the final stretch of broad gauge was converted to standard gauge in 1892. Broad gauge was no more.

Didcot Railway Centre, in the heart of the original GWR area, has a length of broad-gauge track that has been relaid. I tried to get my wife to lie down – silent-movie style – across the rails so I could get a photo with scale, but for some reason she declined.

Didcot has two broad-gauge engines, both replicas. The first, *Iron Duke*, was built in 1985 and is based on the original Iron Duke Class 4-4-2 (not to be confused with the later Iron Duke Class built later and run until 1892) and was constructed from two Hunslet Austerity Saddle Tank locomotives from the 1940s and 1950s. It ran at Didcot and appeared at other heritage railways when in running order. It is now a restored static display at Didcot. The other more recent broad-gauge engine is *Firefly*, which was constructed more recently in 2005. It is a 2-2-2 Class and was built by the Firefly Trust at Didcot. The project, although completed in 2005, was launched in the mid-1980s and the construction plate dates from the late 1980s as the project was gathering pace.

When writing fiction authors try to avoid coincidence, but in real life coincidences happen. In the early 2000s a schoolboy, John Eastwood, visited Didcot on a family holiday and watched the boiler for *Firefly* arrive by low-loader lorry from its manufacturer Israel Newton & Sons and took some photographs on an early digital camera, which I have included with his permission. The coincidence? John Eastwood is now the Managing Director of Israel Newton & Sons, although at the time he had no connection with the company.

I visited Newton & Sons to get more information on the boiler and manufacturing techniques as they still do a lot of work for heritage railways. In the mid-twentieth century, Newton & Sons were working on industrial pressure vessels but in the 1970s, as steam preservation was becoming more common and the last steam engines to be retired started to need work, they found themselves with a new direction. John showed me the hydraulic press used to bend fireboxes into shape and the welding techniques used.

I spoke to John and asked him how his current ownership came about. It was sold by the Newton family to new owners, who John started working for around ten years ago. Two years ago he bought the business himself when they wanted to concentrate on their other manufacturing business. Newton & Sons build and refurbish boilers and fireboxes from all over Europe, at the speed of around five a year. Other smaller projects fill in the gaps in the schedule.

John's schoolboy fascination with heritage railways has led to his ownership of the company that built the boiler he saw being unloaded in the early 2000s. He also volunteers at Foxfield Railway, where his expertise is often called upon.

A lot of heritage railways have turntables often reclaimed and moved from elsewhere. Their use negates the need of large loops of track to turn locomotives round where room is often restricted. At Didcot an example of a turntable exists that is possibly unique, a broad-gauge turntable. Most surviving turntables at heritage railways are standard gauge. Installed as an addition to other broad-gauge exhibits, the broad-gauge turntable came from Devonport Royal Dockyard, built in 1868. It was originally used mainly for wagons rather than larger engines, but it can turn small engines if they fit within its length. Most small ones were powered by simply pushing the carriage at the side to move them; larger ones may have been pulled by a horse or used vacuum motors driven by the locomotive being turned. Didcot also has a standard-gauge turntable in regular use.

The two replica broad-gauge engines are currently non-operational, but another replica, or perhaps recreation is a better term, is in use regularly and was running on the day of my visit to Didcot.

No. 2999 *Lady of Legend* was built to give the Saint Class a presence in the heritage railway scene. The class was originally built between 1902 and 1913 with seventy-seven examples being produced. In 1902 George Jackson Churchward became Locomotive Superintendent at GWR and brought with him some radical new ideas about design, having spent some time in America studying larger locomotives there. Where before one or two sets of driving wheels had been sufficient, with the advent of heavier trains incorporating dining cars and toilets, Churchward designed the Saint Class with three sets.

Brass and wood embellishments were abandoned for a more austere look, gone were the side skirts that hid pistons, and the working components were now in full view. Other railways noticed the innovations and started to copy the ideas.

By 1953 all the 2900 Class had been scrapped. None were saved for future use in preservation as the concept was unheard of at that time. Didcot decided they wanted an example of a 2900 in their collection so in the early 1970s a later development of the class, a 4900, was purchased with a view to building the recreation. Interestingly the original prototype of the 4900 was a re-engineered 2900 so what was planned was a reversal of the original building of the prototype.

The build got off to a slow start, with many people believing it to be beyond the capabilities of a heritage railway workshop, but other major projects including the restoration of No. 6023 *King Edward II* gave the team the confidence to start the process. The build finally got underway in 1995 but really took off in 2009. New driving wheels were manufactured as 2900 sported 6 feet 8½ driving wheels whereas the donor 4900 locomotive *Maindy Hall* ran smaller 6-feet diameter driving wheels. When completed it was numbered numerically on from the last original 2900 (which was No. 2998 *Ernest Cunard*), which had left the works as a new locomotive in 1913. No. 2999 *Lady of Legend* became operational in 2019, 106 years later. Parts of original 2900 Class locomotives were used in the build, a connecting rod from No. 2906 *Lady of Lynn* and the whistle from No. 2910 *Lady of Shalott* among the reclaimed parts used.

A slight deviation from the original 2900 Class is a reduction in height of 3 inches to allow its possible use on main-line railways. Network Rail's allowed height is 13 feet 1 inch and original 2900s stood at 13 feet 3½ inches. Network Rail's height limit is for clearance of power lines. No. 2999 is painted in Brunswick Green, a dark-green colour. One of my previous haunts in my train-riding days was a pub near the original Derby loco works that bears the name 'The Brunswick'. I wonder if there is any connection with the colour of No. 2999?

While researching colliery engines for another project, I came across a locomotive, No. 1340 *Trojan*. When I found that it was restored and in operation at Didcot I was keen to see it along with the others engines listed above, as it has a local link to my own area around a hundred miles away from Didcot.

No. 1340 *Trojan* was built by Avonside Engine Company of Bristol in 1897 for Dunn and Shute of Newport. In 1903 it was sold to Alexandra Docks, also in Newport. During this time the locomotive was unnumbered. The docks had a hundred miles of dock sidings and a passenger line of around 10 miles.

GWR took over the Alexandra Docks Company in 1923. It was at this time it acquired the number 1340. Working around the GWR region, it was based mainly around Cardiff but did find itself as far away as Oswestry. It was sold to Netherseal Colliery in 1932, who used it for fifteen years through a change of ownership in 1942 and nationalisation in January 1947. The mine was closed in December 1947 due to flooding issues. This coalfield in South Derbyshire was close to the Ashby area coalfields mentioned in the Battlefield Line (Shackerstone) chapter and close to my hometown of Burton-on-Trent. No. 1340 was sold on to Alders of Tamworth in 1947. It was sold again to Didcot in April 1968. A lengthy restoration took place including a replacement boiler. It ran in preservation for the first time in 2002. On the expiry of the boiler certificate in 2011 it was again overhauled and returned to active service in March 2021 and is now part of the regular fleet.

At Didcot they have a perfectly preserved panel board. In the early 1960s panel boards came in to use. An interactive signal display, they allowed control of larger areas of track with a reduced number of staff and less equipment, therefore lower maintenance costs. Made of 40 mm square tiles known as dominoes, each one has a small section of the layout on it, such as a signal or a set of points on it.

The panel board at Swindon was decommissioned in 2016. It had been in use since 1968 – it is the same age as the author. It was carefully dismantled and moved to Didcot. Nowadays computer screens have replaced LEDs and fibre optics across the United Kingdom.

Visitors at Didcot can press buttons that previously controlled trains for nearly five decades. Now connected to an interactive simulator rather than actual clanking points and clicking signals, the board shows the layout and trains of the mid-1980s.

The panel board at Didcot is an excellent example of preservation, taken straight out of use and into a museum and reinstalled while technicians who know the mechanics of it are still around to put it back together and spares are readily available, rather than sitting in storage for a decade or so and deteriorating then being put together like a big jigsaw but without the picture on a box to help.

Another piece of railway history is also at Didcot in the form of three lengths of 22-inch diameter pipe, which formed part of Brunel's atmospheric propulsion system. Better known as 'Brunel's Atmospheric Caper', if it had gone ahead pumping stations would have to be built at intervals along the line. It was meant to let trains travel up steeper inclines than the locomotives of the day. It was a sort of 'Steampunk' hyperloop, a system that today is being developed by various firms promising ultrafast trains. The Victorian system failed due to the seals becoming brittle and breaking. It is an often-repeated anecdote that if the seals were kept soft with tallow, then the rats ate them, although there's no mention of this in minutes of meetings that took place at the time, it may have happened but wasn't recorded for the archives. Perhaps the system had more significant problems than rats. Several short lines did operate but with limited success. Brunel, impressed with the power and the silent running, felt he could iron out the problems and started building a short line. The pipes were recovered from Devon for display at Didcot. His experimental line ran between Exeter and Newton Abbot and was only in operation seven months. It was more than the price of a steam railway to run per mile at three shillings and one pence (compared with one shilling and four pence for steam).

Didcot also has the only surviving coal stage. Wagons loaded with coal unload at the top of a steep incline and fill the top of the building. Locomotives are brought underneath and loaded with coal. The building had just been renovated and the scaffolding was coming down at the time of our visit.

Didcot Railway Centre is accessed via the current Didcot Parkway railway station.

Broad-gauge turntable.

The *Iron Duke* broad-gauge engine.

Firefly broad-gauge engine.

Arrival of *Firefly* boiler from Israel Newton & Sons. (© John Eastwood)

No. 2999 *Lady of Legend*.

Trojan. (© Didcot Railway Centre)

Didcot's coal stage – believed to be the only example preserved for use in heritage railways.

Signal control panel, an interactive exhibit – relocated from Swindon when it was decommissioned.

Propulsion tube remains of 'Brunel's Atmospheric Caper'.

An air-raid shelter at Didcot provided for railway staff during the Second World War.

'I have to tell you now that no such undertaking has taken place and that consequently this country is now at war with Germany.' Neville Chamberlain's speech was received on sets like this across Britain in 1939.

Chapter 6

Great Central Railway: Loughborough to Leicester

Great Central Railway does nostalgia excellently. Heritage railways are always about keeping the past alive. Whether it's steam or diesel, GCR take it to the next level. The set dressing is unparalleled, a mock-up of a 1940 W. H. Smith & Son sells snacks and drinks while displaying vintage *Punch* and *Practical Mechanics* magazines as though they had just been delivered and were the current issue, the staff in period clothes. The station is presented in early 1950s style and although the film *Brief Encounter* was made in 1945 and filmed mostly at Carnforth station in Lancashire, it felt like I had wandered on to the set of it – I almost expected Celia Johnson and Trevor Howard to sneak out of the waiting room behind me.

GCR operates between Loughborough and Leicester and is virtually unique in that rather than a branch line, it actually operates on a former main line, uses four stations and is currently the only double-track main-line heritage railway in the world with over 5 miles of working double track. The current Network Rail main line in use over the route actually predates GCR's double-track main line having been built around 1840. GCR's track was built in 1899 by a separate company and was the last main-line track to be laid unless HS1 is included.

On the day I visited, steam was running in the form of No. 73156, a Standard Class 5 locomotive (not to be confused with the earlier Black Five Class designed by William Stanier and built between 1934 and 1951). No. 73156 is one of a 172 examples built between 1951 and 1957 by British Railways. The design work for the new Standard Class 5 was coordinated by Doncaster but the bulk of production came out of Derby, 130, the remaining 42 coming out of Doncaster. The last one was built in the summer of 1957. No. 73156, the GCR's example, was built at Doncaster in 1956. They could reach just under 100 miles per hour. They were used on local express routes around the Midlands, Sheffield and Leeds, along with longer express routes, London–Southampton–Bournemouth–Weymouth. They were in operation right up until 3 August 1968, the last day of scheduled steam traction in the UK. (The last actual day of steam was a week later, on 11 August, with a special steam tour known as The Fifteen Guinea Special. The price was the result of high demand, and the ticket price in today's money would be £350.)

GCR's example, No. 73156, was sold for scrap after just eleven years and 325,000 miles, arriving at Woodham Brothers in Barry, South Wales. Of the 172 built, just 5 survived into preservation. No. 73156 was purchased by a newly formed group, The North West Locomotive Action Group – later this group became Bolton Steam Locomotive Co. After a few years in Bury slow progress was being made and some of the time in storage was outside, the precious engine shed space given to projects further advanced so a deal was struck with GCR to provide workshop facilities in return for a lease for use on the railway when complete.

Like Peak Rail's Class 44 engine – D8 *Penyghent* – GCR was lucky that their locomotive No. 73156 also spent a lot of its life working on the track that it now runs on in preservation. It was allocated to Neasden, which was the engine shed in North West London for GCR trains and regularly worked to Woodford Halse, Rugby, Leicester, Nottingham and Sheffield, with occasional runs through to Manchester. It was transferred away from Neasden in April 1958 to sheds in Sheffield, then Derby, before returning to Neasden in September 1960. For six months from June 1962, it became a Leicester engine. It left GCR in May 1963 first for Cricklewood before ending up in Bolton in 1966, two years before being withdrawn.

One of the major problems for the restoration was the lack of its tender (the unit pulled directly behind the engine carrying coal and water). It was sold off separately while in Barry. No suitable tenders, a BR1B type, were available for purchase so a new one was built from scratch. GCR believe this to be the only time this has been done for a preserved locomotive.

No. 73156 made its first run in preservation on 5 October 2017, almost fifty years to the day (24 October 1967) since it was withdrawn from service in Bolton.

Another steam engine also works at GCR: No. 92214, a Standard Class 9F, which was built in 1959. It is a recent addition to the manifest at GCR, being bought after its appearance at the 2014 winter steam gala. It is believed to be the last 9F to enter service and was withdrawn just six years later in 1965. It spent time at Peak Railway before ending up at GCR. In the past it has carried different names including *Cock o' the North* and *Central Star*.

An early Railbus unit on duty during our visit was a Class 101 set, two twin-engine railcars. The one photographed is DMCL M50321 and the other unit out of sight further down the platform is DMBS E50321. They are predecessors of the Pacer Class of locomotive like the 141 and 142 on duty on my visit to Midland Railway – Butterley. The 101 Railbus Class was built between 1956 and 1960.

At Quorn station on the line is a NAAFI (Navy Army and Air Force Institute) style tearoom. Like Loughborough, it also does excellent set dressing, looking just like it would have done in the 1940s, including price lists and warning signs about keeping your mouth shut.

The mock-up of a NAAFI canteen shows an insight to the real NAAFI organisation formed in 1920 by the British army by amalgamating Expeditionary Force Canteens and the Navy and Army Canteen Board. It supported the armed forces personnel both at home and abroad. When Britain's forces went to war in 1939 the NAAFI followed them, often into hostile environments. There were 10,000 NAAFIs in the Second World War with 110,000 personnel, 550 of whom lost their lives during the 6 years of hostilities.

Close to Quorn station is another link to Britain's war effort: Beaumanor Hall. In 1939 the hall was requisitioned by the War Office and used as Number 6 Intelligence school and then became one of the most important 'Y Stations' in the country, a listening post intercepting coded German messages that were then passed on to Bletchley Park for breaking often transported via motorcycle couriers. An elderly family friend, 'Bunty', now sadly passed away, was a motorcycle courier and often regaled us with tales of her late-night missions the length and breadth of the country, but wouldn't go into details, even in her nineties – the Official Secrets Act signed during her service was not broken by her. I often wondered if she carried intercepted messages to Bletchley perhaps from Beaumanor Hall. Hundreds of military personnel working at Beaumanor Hall would also have passed through Quorn station.

Radio fingerprinting was pioneered at Beaumanor in room 61 on the top floor, detecting where radio messages originate from by analysing the transmission characteristics of the original transmitter. Over the years this has been developed into detecting mobile-phone transmissions and is used by both police and military around the world.

The first confirmation of the success of Operation *Chastise* (the Dambuster Raids) was received by Beaumanor Hall and passed on, a very important contribution to Britain's war effort close to GCR. The hall is now a conference centre and wedding venue, but history tours are available and can be booked.

The other two stations on the line are Rothley and Leicester North. Rather than recreating the 1940s, Rothley is dressed as it would have been in George V's era, specifically 1912. The station provided important transport links to the village. As well as passenger services, goods were brought into the area, milk, farm machinery and general parcels arriving at the station for local residents.

Restored gaslights provide the lighting for all the buildings and the platform (when required) at Rothley, creating a unique nostalgic atmosphere. Electricity was never installed at the station during its original operating life and only got power when GCR heritage railway started operating the station in the 1970s. Heating in the tearoom is provided by a roaring fire.

Rothley station is of the standard design of stations on the 'London Extension' of the Manchester, Sheffield & Lincolnshire Railway. An island platform served two lines accessed from a stairway from the road crossing the line.

Leicester North is the final station on the line. It replaces Belgrave and Birstall station, which originally opened in 1899 but was closed by British Rail in 1963. It spent the next few years being neglected and vandalised, and it was decided to demolish it in 1977. The new station was built a little south of the original and renamed Leicester North, originally consisting of just a platform. It was opened in 1991, with several guest locos including Stephenson's *Rocket* replica from the National Railway Museum at York. The station buildings were added later, officially being opened in 2002. The line from Rothley is a single-track section. The two platforms were built with the idea of adding a terminus at the end of the line, but this plan never came to fruition. A canopy was added in 2009 and a tearoom, The Tail Lamp Tea Room, offers homemade cakes and refreshments.

I came across two railway terms while researching GCR: 'Runners' and 'Windcutters'. They both mean the same thing, The first is the preferred and (possibly) original term used by the drivers, while the second is a more recent term originating in the preservation railway era creating a more dramatic term. Trains of coal supplies, incorporating often as many as sixty wagons long, would travel down the less-used main line of the original GCR route. The trains required long stopping distances, and a double signalling system was used with outer and inner distant signals. If the outer signal was green, the route was clear all the way through and trains could therefore travel through the stations unchecked, hence 'Runner'. The name 'Windcutter' though has a more romantic feel about it.

A retro W. H. Smith & Son.

Platform at Loughborough.

Period platform memorabilia at Loughborough.

Locomotive No. 92214 *Leicester City*.

Locomotive No. 73156, GCR's Standard Class 5.

Locomotive No. 73156, GCR's Standard Class 5.

Class 101 Railcar M50321.

The NAAFI at Quorn station.

Memorabilia at Loughborough station.

Chapter 7

Severn Valley Railway: Kidderminster

The Severn Valley Railway is one the most famous heritage railways in the United Kingdom, having been used in films and television for decades. *God's Wonderful Railway* was one of the first television programmes I remember watching about steam railways. Filmed in 1980, it was a children's drama series featuring a station and a fictional family, 'The Grants', on the Great Western Railway. It was split into three historical periods: Victorian times covering the building and opening, Edwardian period of operation and finally the Second World War. I've not watched it since but have vague memories of it and remember it being engaging and well made. Two years prior to this being made, Robert Powell as Richard Hannay hung off one of the bridges on the line in the 1978 adaptation of *The Thirty-nine Steps*. Add in *The Chronicles of Narnia* and *Sherlock Holmes: Game of Shadows* and Severn Valley Railway's film pedigree becomes impressive.

The railway was built between 1858 and 1862 by the Severn Valley Railway Company (not connected to the present company operating the heritage line). It opened in February 1862 and was operated by the West Midland Railway (WMR), who ran it for a little over a year, until being absorbed into the Great Western Railway in August 1863. The line ran from Hartlebury (near Droitwich) and Shrewsbury, a distance of around 40 miles. The principal original stations were Stourport on Severn, Bewdley, Arley, Highley, Hampton Loade, Bridgnorth, Coalport, Ironbridge/Broseley, Buildwas, Cressage and Berrington. A link was added in 1878 to join Bewdley and Kidderminster. Surprisingly the Severn Valley line was not a victim of Beeching's cuts laid out in his 1963 report 'The Reshaping of British Railways' as the line's future had been decided the previous year, being deemed unprofitable.

In 1965 a group of rail enthusiasts met in the Kidderminster pub the Coopers Arms, offering £25,000 for an approximately 5-mile section of the line that had closed in 1963. A new company, Severn Valley Railway Company, was formed, a light railway order was obtained, and a deposit was paid to British Rail for the track. The group's intention was stated at the time to be 'preserve, maintain and restore the standard gauge railway extending from Bridgnorth to Kidderminster via Bewdley'.

The service started in the summer of 1970 between Bridgnorth and Hampton Loade. The balance of £25,000 for the line was paid to British Rail. A further 8½ miles of track were added having been purchased in 1969 and finally in 1982, upon the closure of the sugar factory at Foley Park, the final length of line to Kidderminster was obtained. Services to Kidderminster started in July 1984.

There are two museums on the line, both are free to Severn Valley Railway passengers. A small memorabilia museum is at Kidderminster housing a signal-box frame and a collection of phone-network switchboards. Nowadays modern phone systems can handle multiple lines and extensions with very little intervention from employees. In the early days of telecommunications small switchboards were used by organisations usually operated by a receptionist doubling up as the operator. Railway networks would have their own switchboard along with large factories. It's interesting to see a large manual machine that did the job now done by a silicon chip in the phone itself. The museum to one side of the platform at Kidderminster is well worth a look inside.

Once again set dressing is done well on SVR, another W. H. Smith & Son mock-up similar to the one at Great Central Railway again selling modern refreshments alongside nostalgic notices. One of the newspaper posters declares 'Fred Perry wins Wimbledon'. A gift shop complements the vintage newsagents.

Steam and diesel were both running on alternate timetables the day I visited. The steam locomotive was No. 75069, which is a Standard Class Four locomotive. The class entered service in 1951 and was designed for mixed traffic use on secondary lines where Standard Five's and its predecessor the 'Black Five' would have been too heavy.

There was a total of eighty Standard Class Fours built between 1951 and 1957. Six have survived into preservation. No. 75069 was rescued from a scrapyard in Wales. Some of the class were fitted with a double chimney (or double stack). From the outside the appearance is similar; there is no physical second chimney but rather an elongated oval chimney with the two separate holes below, both exhausting into the same chimney. The longer chimney can be seen in the photograph of No. 75069. A single chimney example would have a round chimney. Both types have been saved into preservation.

The rolling stock consisted of Mark 1 open and compartment carriages. I couldn't resist the compartment option, sharing it with a couple taking their grandchild on the railway during the school holidays. A steam locomotive pulling Mark 1 carriages, how Harry Potter is that?

The diesel engine on duty was L142 *Sir John Betjeman* resplendent in London Underground maroon livery. L142 is a Class 20, as described earlier with another in use at Shackerstone, with the nickname of 'Choppers'. I managed to grab photographs of *Sir John* but rode back to Kidderminster behind No. 75069.

At Highley is the Engine House Museum, an interactive centre where I could have spent a week examining all the exhibits. There is a recreation of the Coopers Arms where the preservation of the line began its long journey back in 1965, and an enormous Thomas the Tank Engine train set took up the middle of the hall with twenty or thirty children playing with it all at once.

Linked to Thomas the Tank Engine, the Engine House is also home to a full-size *Gordon* resplendent in blue. *Gordon* wasn't named in honour of Reverend Awdry's creation but Gordon of Khartoum, though this didn't stop the name plate being stolen and later being put up for auction. The auctioneer, a keen rail enthusiast, spotted the significance of the lot and informed the police.

Originally built quickly for limited use during the Second World War, *Gordon* survived the war and went on driver training duty and had a top-secret job running supply trains during the Suez Crisis in 1957. Upon retirement in 1971 it joined the Severn Valley fleet, where it ran until a boiler failure in 1998.

Standard Class Four Tank Engine No. 80079 (not to be confused with the Standard Class Four mentioned earlier in the chapter) has a sad story to tell. On 30 January 1958 in thick fog it ran into the back of another train near Dagenham, killing ten passengers and injuring a further eighty-nine others. The rear carriage of the other train was a 'ladies only' carriage and was said to have 'splintered like matchwood' by eyewitnesses. As several factors led to the crash and no single person or event was deemed the cause, no compensation was paid to the families of the deceased or the injured.

Two interesting exhibits in the Engine House are carriages with significant histories. The older of the two was King George VI's coach built in 1942, which is one of two coaches built for the king and queen (later known as the Queen Mother). The queen's is currently at York Railway Museum. The two separate coaches met at a 'reception' where double carriage doors allowed the royals to alight side by side. The accommodation consisted of (from the reception backwards) saloon, king's bedroom, bathroom, equerry's bedroom, then equerry's office and second reception. An early air conditioning system was fitted. After the king's death it was later used by Prince Philip and sometimes the then Prince Charles (King Charles III now). It was replaced in 1977 for the Queen's Silver Jubilee.

During its service it travelled thousands of miles with royalty on board, especially during wartime when the king visited towns and cities that had been bombed. For me, I like to find the tiny details with a bigger story to tell. In the Second World War the public were told to use just 6 inches of water for a bath, water rationing as it were, and it was well known that the king also adhered to this request. The bath has a faint red line still visible indicating a 6-inch depth. I was also impressed by the relative starkness of the carriage. While well built, it was very austere, not flaunting wealth to the general public at a time of rationing and shortages. The glass is bulletproof and the carriage is armour plated in case of aerial bomb attack or sabotage. An extra wheel was added to the bogie to give a more comfortable ride and possibly take the extra weight as the carriage is one of the heaviest ever built. Its movements were a closely guarded secret, with the crew only being told a few hours before any journeys were made. Across the walkway is another carriage with an equally compelling but more sinister story.

In the early hours of 8 August 1963, a gang of fifteen men tampered with the railway signals at Bridego Railway Bridge (now named Mentmore Bridge) at Ledburn. The green light on the signal was covered so it couldn't be seen and an external battery was used to illuminate the red signal, to create a false stop signal. The objective of the tampering was to stop a train. An unnamed sixteenth man, a retired train driver, was thought to be present. The train in question was carrying £2.6 million in cash (the current value, allowing for inflation, would be over £60 million.) While no firearms were used, the driver was coshed over the head and suffered severe injuries. He returned to work after the robbery but never got over the trauma, retiring in 1967 but sadly died soon after in 1970.

The retired driver who was recruited by the gang to move the train was not familiar with the locomotive used, as he had previously been employed driving shunters.

The gang was traced to a nearby farmhouse and although they had fled, vital fingerprints were left behind that led to the capture of most of the members. Less than £400,000 was recovered, the last of which (£40,000) was found in a phone box in Newington, South London.

The coach in the engine house is the Mobile Sorting Office and is a similar coach to the ones in use that night. As well as the history of the robbery displayed on boards, a BBC news report plays on a screen (along with an original promotional film of *Night Mail* by W. H. Auden). It is also set out as an interactive learning exhibit where children of all ages can help to sort post. There is a single surviving Mobile Post Coach from the actual train involved in the robbery, which is at Nene Valley Railway.

Just outside the Engine House parked in a siding was a 'locomotive' that caught my attention, now sadly a recent victim of graffiti (an archive image of just months before my visit showed it un-vandalised). It was a small Railbus type that I was unfamiliar with, but after some research I found that it was the prototype of the Parry People Mover. Designed by John Parry in the early 2000s, its designated type is PPM 50 unit and it was originally numbered 999 990. It was then renumbered 139 000 before finally ending up numbered 139 012 that it carries now. The Parry People Mover model that is currently in use on the Stourbridge Junction to Stourbridge line is the PPM 60. Only two were produced, which are slightly longer than this prototype. The Stourbridge link line at 0.8 miles claims to be the shortest rail network line in Europe, taking just three minutes and costing £1.40 for a single ticket (at the time of writing).

On my return to Kidderminster to pick up the car, another locomotive was now in view on the line, a Class 42 in fresh paint following its recent refurbishment. The Class 42 is also known as 'the Warship' Class and is a diesel-hydraulic locomotive. The 42 at SVR is D821 *Greyhound*, built in 1960. It is owned by 'the Diesel Traction Group' (DTG). The class is a revised design of a post-war (West) German locomotive, a scaled down version of GFR (German Federal Railways) V200 and is powered like its German relation by Maybach Diesel engines built under licence by Bristol Siddeley at their factory near Coventry. The official design speed was 90 mph but due to the hydraulic transmission, which could not be governed precisely, the locomotive often recorded running speeds of up to 100 mph.

D821 was purchased for preservation and ran in preservation for the first time in 1973. A different engine was originally to be purchased by DTG, Class 22 (Baby Warship) D6319, but was scrapped by accident at the Swindon works, so several alternative locos were offered. D821 was chosen as it was in the best mechanical condition of the offered engines. Only two of the class now remain in use: D821 at SVR and D832 *Onslaught* at East Lancashire Railway.

The stations on the Severn Valley Railway in order from Bridgnorth to Kidderminster are Bridgnorth, Eardington, Hampton Loade, Country Park Halt (request stop), Highley, Arley, Northwood Halt (request stop), Bewdley, then terminating at Kidderminster. As well as the big three stations (Bridgnorth, Kidderminster and Highley), Hampton Loade is described as 'a station frozen in time' and plays a significant role in the 1940s weekends often held on the railway. Light refreshments are sold here most weekends (along with larger outlets at the other stations) and the River Severn is a short walk away. Traffic is very light around Hampton Loade as vehicle access is down a very long and narrow country lane. The nearby Unicorn Inn serves food and beverages. As well as the bigger stops on the line, the station and hamlet of Hampton Loade is very much worth getting off for and having a look round if just for the perfect tranquillity and perhaps a pint at a village pub.

Kidderminster station (interior).

Another retro W. H. Smith & Son, this time at Kidderminster.

65

Vintage suitcases as set dressing on the SVR.

George VI's royal coach sitting room.

Royal coach bathroom.

The faint line still visible on King George VI's bath indicating a 6-inch depth due to wartime water rationing.

Mobile sorting office coach.

Gordon nameplate that was once stolen.

Locomotive No. 75069, a Standard Class Four fitted with a 'double stack' chimney (note the oval profile).

Parry People Mover prototype 139012.

Locomotive D821, a Class 42 locomotive based on a German design.

Chapter 8

Keighley and Worth Valley Railway: West Yorkshire

'Flannel Petticoats!' or perhaps you prefer the more tear-jerking quote of 'Daddy, my Daddy!' Both are from *The Railway Children*. The film was shot on the Keighley & Worth Valley Railway (KWVR) in 1970. Directed by Lionel Jeffries, it starred Jenny Agutter, Sally Thomsett and Gary Warren as the Waterbury children, Dinah Sheridan as their mother and Bernard Cribbins giving perhaps his finest performance of a long and illustrious career as the stationmaster. It is often credited as being the perfect children's film. A remake and a sequel, while good, didn't quite reach the high point of the original film. As well as *The Railway Children*, some filming of *Peaky Blinders* took place on the line and the recent adaptation of *All Creatures Great and Small* used it extensively.

This book is predominately about heritage railways in the Midlands, where there seems to be quite a large number of them, but having already stretched the Midlands a little south to Didcot to include it in this book because of my passion for Brunel and broad gauge, I'm stretching it northwards to West Yorkshire to include Haworth and the KWVR.

The branch line's existence is in part due to the Brontë sisters. In 1861 civil engineer John McLandsborough visited Haworth to pay his respects to the Brontë sisters and in particular Charlotte, who had been the last sister to die in 1855, six years earlier. The engineer was surprised that Haworth was not yet served by a railway, so suggested the idea.

The Act of Parliament incorporating the railway was issued early in 1862, John McLandsborough was appointed as the engineer and work began on Shrove Tuesday (9 February) 1864. John's civil engineering experience was mainly in water and sewage works but he had limited experience of railways, having been involved in the building of the Otley & Ilkley Railway. Another engineer, J. S. Crossley from the Midland Railway, was appointed as a consultant engineer. The original timescale was one year but delays due to ground problems, in particular quicksand at the tunnel site near Ingrow West, required further piles to be inserted into the bedrock, which in turn damaged the foundations to a local church, which led to a repair bill of nearly £2,000. The line was finally finished in 1866. The track was laid from both ends, meeting in approximately the middle of the line.

The opening ceremony was held on Saturday 13 April 1867, with the regular passenger train service beginning on Monday 15 April 1867. The line was operated by the Midland Railway in part due to them wanting to protect their region from the Great Northern.

Following the Railway Act of 1921, which required the amalgamation of over a 120 regional railways into just 4, it became part of the London, Midland & Scottish Railways (LMS) in 1923. Later, in 1948, it became part of the nationalised British Rail.

The line ran its last British Rail passenger service on 30 December 1962 and on 18 June 1963 freight traffic stopped. Five days later, 23 June, a chartered train ran on the line from Bradford to Oxenhope and back, organised by the newly formed Keighley & Worth Valley Railway Preservation Society. Soon after this excursion, the line closed for several years. It reopened on 29 June 1968 as a preservation railway. The railway was used twice for *The Railway Children*, a seven-part BBC adaptation released in 1968, and the more famous feature film two years later. Bobby (Roberta) Waterbury was played in both by Jenny Agutter, and over fifty years later she reprised her role in *Return of The Railway Children*, once again filmed on the line.

On a previous visit my wife asked to be photographed in the Carriage Works Museum at Ingrow West station, next to the cut out of the *The Railway Children*. 'Why?' was my answer. 'Because X (I won't reveal which one) might be related to me.' I found out that day my wife was possibly related to a railway child as a very distant cousin. Another cousin found out years before I met my wife while doing their family tree. How we had been together for ten years and married for seven before I found out this information I'll never know, but I should have had an inkling when she removed her petticoat on an early date and waved it at a train (I might have made the last line up).

Most of the railways featured in this book are branch lines, therefore go through quaint villages that were not deemed profitable as motoring became more accessible to the general public in the post-war years, so most are very photogenic and pretty stations. However, not to take anything away from the other railways in this book, KWVR does seem to have more than its fair share of stunning views and interesting villages.

We joined the railway on our visit part way down the line at Ingrow West and travelled all the way to Oxenhope, where the exhibition shed is located (not to be confused with the Rail Story also on the line at Ingrow West). In the exhibition shed was an engine I wanted to see, which was No. 118 *Brussels*, a Huntley Austerity Saddle Tank. In 1942 the War Department started to stockpile small railway engines for the forthcoming invasion of France. Needing to be built quickly, initially the LMS 3F locomotive was to be the standard design but manpower resources showed that a simpler locomotive was required. Hunslet stepped in and convinced the government that their basic shunter design would be a better fit for their needs. The 50550 Class, built in 1941, which was an enlarged version of a design from 1923, was simplified into the Austerity Saddle Tank engine. The work was contracted out to various manufacturers to meet demand for D-Day. After the war, production continued as they turned out to be an excellent workhorse. The Coal Board ordered seventy-seven engines, which were delivered between 1948 and 1964. The British Army ordered an additional fourteen examples to supplement their existing ninety left over from the Second World War. KWVR's example, No. 118 Brussels (formerly 71505), was built in 1945 and was used at Longmoor Military Railway – the army's main training railway. Twenty-eight examples were used or stored there until its closure in 1970 and this example was sold to KWVR in 1971. It is currently set up as an oil-burning engine rather than coal, which is one of the reasons that it is currently not on their running engine rota. Another reason is that it is too small for regular use so it remains in the exhibition shed as a static display.

The reason for my interest is that *Iron Duke* at Didcot, the first of their broad-gauge replicas to be built, was built from two Austerity saddle-tank engines. I wanted to see how *Iron Duke* had started life.

Also in the exhibition that's worth a mention is No. 51218, which was built in Horwich in 1901. It remained in use until 1964. It was bought by members of the Roche Valley Railway Society and moved to Haworth in 1965. After being retubed in 1974 it ran until the 1990s, getting its first major overhaul since joining the KWVR. It returned to the track in 1997. It last ran in 2006 and requires another overhaul but was cosmetically restored for display in 2018.

After a whistle stop tour of Oxenhope and the exhibition shed we got back on the same train and went back to Haworth.

Haworth is the star of the line for me. A visit to the Brontë Parsonage Museum may be of interest as it gives a unique insight into the family's literary life. The cobbled high street is famous and was part of the Tour de France route, when it started in Yorkshire a few years ago. I struggled to walk up it, let alone ride my bike up it. Tea shops, antique shops and some amazing gift shops are all worth a peek inside.

A stop at Oakworth had to be done. *The Railway Children's* home station, you can follow a Railway Children's trail taking in many locations of the film. The tunnel just outside the village is Mytholmes Tunnel, the location of Jim's injury in the film. The station looks very similar to how it did in the film and it felt like Bernard Cribbins could pop out of the ticket office at any moment.

The other end of the line from Oxenhope is Keighley. KWVR share the station with the main-line service, but from its own platform and ticket office.

Back to Ingrow West and a visit to the two museums at the station, the Engine Shed and the Carriage Works. The Engine Shed shows the building of a steam engine from the drawing board, through the building to testing and use, with several exhibits to tell the story including original drawings and locomotives and a rather strange-looking stationmaster dummy. The Carriage Works holds one of the largest collections of historically important rolling stock in the country including 'The Old Gentleman's Coach' from *The Railway Children*. It occasionally gets used on very special occasions. While on the station site, the two museums are run by different organisations to the KWVR. The Carriage Works is owned and operated by the Vintage Carriages Trust and the Engine Shed by the Bahamas Locomotive Society.

On duty the day of our visit were two different steam locomotives. The first one we were pulled by was No. 75078, another Standard Four as previously described in the Severn Valley Railway chapter. Again, like No. 75069, it was a double-chimney example with the oval chimney. It was saved from the same scrapyard in Barry, South Wales, as No. 75069. Built in 1956 and withdrawn ten years later, it was saved by KWVR in 1972. Restoration took five years and it returned to service in 1977. Over the years several further withdrawals and overhauls have taken place, the most recent of which was finished in November 2022. This locomotive took us to Oxenhope and back to Haworth.

The other train in service the day of our visit has a very interesting story, as it was *The Green Dragon* in *The Railway Children* (film).

Officially known as a Lancashire & Yorkshire (L&Y) Railway Class 25, it was built in 1887. There were 280 examples made by several different private companies to a W. Barton Wright design of 1876. KWVR's example was the last unit built by Beyer Peacock of Manchester and was given the original number of 957. The Class 25 was superseded by the L&Y Class 27 designed by John Aspinall and built in house by L&Y Railway at their new Horwich works (the L&Y Class 27 is not to be confused with the later BR Class 27 Diesel unit). No. 957 became part of the LMS fleet in 1923 and was given the number 12044. Later in 1959, while numbered 52044, it was retired by its then owner British Rail, after seventy-two years of network running.

It was rescued by Tony Cox, who wanted to preserve an L&Y locomotive and while 120 Class 27s were still in use, only two Class 25s were still being used, so he decided to preserve one of the earlier class. Of the two examples in use, the other one, No. 52016, was in better condition and was the original target for the rescue. Unfortunately it was badly damaged in an accident and was scrapped, leaving No. 52044 the only viable unit. The purchase price was £900 including delivery.

Tony Cox later became secretary of the Worth Valley Railway Preservation Society and No. 52044 followed him there in 1965. Essential remedial work was carried out and it was used in two films released in late 1970. One is obviously *The Railway Children* and the other is Billy Wilder's (the Austrian-born Hollywood director whose earlier films included *Some Like It Hot* starring Marilyn Monroe) *The Private Life Of Sherlock Holmes*. The railway scenes were filmed extensively on the KWVR, the rest on location at Loch Ness involving a monster-shaped submarine.

No. 52044 was running passenger services on the line by 1971. It was withdrawn a few years later in 1975 and left for over two decades. In 1996 the engine refurbishment was outsourced to the Severn Valley Railway's workshop facility and in 2002 was returned to fully working order undergoing trials at SVR during Easter 2002. It was returned to KWVR and spent eleven years in regular use. In 2013 it was once again sidelined, this time in Haworth Shed. Another overhaul was carried out from 2016. In July 2021 successful steam tests took place and it is now in regular use again.

Also in the Carriage Works is a small locomotive that is no longer in use, but appeared on the cover (as well as the three children) of the LP record (a reading of the book by Lionel Jeffries) of *The Railway Children*. The engine Hudswell Clarke No. 402 or *The Lord Mayor* was built in 1893 and used in various construction projects across the country including the Becontree housing estate in Essex, which at the time was the largest housing estate in the world. It also worked in the docks in South Wales. These small engines were designed to be moved around the country by road on low loaders often pulled by steam traction engines. They transported bricks around the sites they were employed on. *The Lord Mayor* is unusual in the fact that it was standard gauge, as most construction engines used narrow gauge to aid transportation. *The Lord Mayor* was retired in 1968 and donated to the charity The Lord Mayor's Trust for preservation and moved to Haworth.

Also in the Carriage Works is other rolling stock used in films including one used on the 2016 adaptation of Arthur Ransome's *Swallows and Amazons*.

When at a heritage railway always try to seek out the various museums, like the Engine Shed and Carriage Works. These are often run by another charity and need small donations and money spent in their gift shops to aid their survival and they all have stories to tell.

Station exterior at Ingrow West.

Locomotive No. 75078 at Ingrow West station.

Locomotive No. 51218, the first engine to arrive at KWVR in 1965.

Locomotive LMR118, a Hunslet Austerity Tank engine (two examples of this type were used in the building of the loco *Iron Duke* at Didcot).

Howarth station exterior.

Locomotive No. 52044 *The Green Dragon*.

'Daddy, my Daddy!' Oakworth station, *The Railway Children*'s station.

Carriage Works Museum at Ingrow West.

78

Locomotive Hudswell Clarke No. 402 or *The Lord Mayor*. Used in promotional images for *The Railway Children*.

Platform at Damems.

79

Chapter 9

Foxfield Railway: Near Stoke, Staffordshire

The original Foxfield Railway started life in 1883 as a narrow-gauge line serving the local colliery. In part it followed the route of an earlier type of railway – a plateway, which was an 'L'-shaped rail that ran horse-drawn wagons. The 1883 narrow-gauge line hauled coal for around ten years until more capacity was required, so a standard-gauge line was built to link the Foxfield Colliery to the North Staffordshire Railway. It was built in 1892–93, at weekends by North Staffordshire Railway's employees. It was built as a private venture, therefore took the shortest route, up steep inclines and sharp bends. It was also built using second-hand materials from ex-North Staffordshire Railway's stock. The line was joined to the Derby–Crewe line. It opened in January 1894.

The short line closed in 1965, but local volunteers created the Foxfield Light Railway Society to reopen the line as a heritage line. When it reopened passengers were carried in converted trucks out of the old colliery site at Dilhorne pulled by a tank engine. In time coaches were purchased and a new station was built at Caverswall Road, Blythe Bridge. The line runs for 2 miles, from the station to Foxfield Bank. In 2009 the line featured in *Return to Cranford* starring Dame Judi Dench.

As I was driving to Foxfield Railway, the big question was, would No. 3 be running? The reason for my interest in this particular locomotive is that it spent its working life less than a mile from my childhood home in Burton-on-Trent. The breweries in Burton used to be the town's main employers (or businesses linked to brewing – my late father worked for Grundy's, a barrel and keg manufacturer). The breweries had a network of railways criss-crossing the town. A level crossing for Bass on High Street was the subject of a painting by L. S. Lowry, one his few 'Matchstick Men' paintings of a subject outside of his hometown of Salford, Manchester.

No. 3 is a Hawthorn Leslie locomotive built in 1924 with original works number 3581. It was delivered new to Marston, Thompson and Evershed, a brewery in Burton-on-Trent, where it remained until its retirement in 1967 (although from 1955 to its retirement, it was the 'spare' engine, used five weeks a year while its replacement was having its annual service).

The small locomotive moved to Foxfield Railway in April 1967 and, together with Henry Cort – now on display in the Foxfield Museum – worked the first passenger train of the preservation era (and first passenger train ever) at Foxfield on 14 May 1967. It then went to GCR in Loughborough in 1970. It ran on the GCR line for three years until being withdrawn for overhaul. After the repairs it ran until 1996 before returning to Foxfield for a full restoration. It began running again late in 2001.

It was running and when I exited from the ticket office onto the platform, I found it preparing for its runs that morning resplendent in black and gold and carrying its *Marston, Thompson and Evershed No. 3* nameplate. Pulling several Mark 1 open carriages, we travelled the length of the line, getting out at the end of the line to stretch our legs before the return journey.

In the museum is one of only a few surviving North Staffordshire Railway 'Signalman's Chairs'. A signalman could be on shift for as long as twelve hours and to allow for short rests chairs were included in signal boxes on the North Staffordshire Railway.

In 1872 a goods train collided with a stationary passenger train at Stoke-on-Trent standing at the Up platform. (In the English Railway network the 'Up platform' is actually the one leading to London and not north as you would think, hence the term 'Up to London' regardless of your geographical location. In Scotland the 'Up platform' usually heads towards Edinburgh unless part of the East or West Coast UK main lines.) The accident report concluded that the Signalman's Chair contributed to the accident by making the signalman too comfortable and therefore less able to concentrate on the job in hand. It didn't blame overwork from twelve-hour shifts. The chairs were all removed, but a few survived including the one on display at Foxfield.

Also in the museum was another locomotive that had a link to my hometown. *Florence No. 2* is an engine that was originally at Florence Colliery in nearby Stoke-on-Trent along with its sister loco (*Florence No. 1*). They were replaced by diesel engines in 1968. Built in 1953 by Bagnalls of Stafford, it worked a steep 2-mile stretch linking the colliery to the main line at Trentham. It was retro fitted with a Giesel injector and flat-sided chimney to reduce emissions, which it still carries. It was one of a class designated 16 inch and was a 0-6-0 configuration and a saddle-tank design. Eighteen were built between 1942 and 1955.

Florence No. 1 was scrapped in 1971 whereas *Florence No. 2* remained the spare engine at Florence Colliery until 1975, when it was moved to Cadley Hill, a colliery between Burton-on-Trent and Swadlincote. Three years later it went to the Battlefield Line for preservation. It was purchased by a Foxfield member in 2000, restored to running condition and used for some years. The museum also has the whistle from *Florence No. 1*.

Other examples of the class have survived into preservation, possibly most notably *Courageous* (previously known as *Birchenwood No. 4*, which spent its working life also near Stoke at Birchenwood Gas and Coke Company), which is regularly run at Ribble Steam Railway (not a railway I have visited yet).

Another locomotive called 'No. 2' is in the museum and which is the last surviving engine from the North Staffordshire Railway, built in 1923. The locomotive is a New L Class. It was renamed *Princess* in 1938 and in 1946 was rebuilt with new pistons, boiler and tanks. In 1964 the tank and the recently replaced pistons, boiler and tanks were put on the frame and wheels of another New L Class locomotive, No. 72 (built in 1920). This locomotive then adopted the No. 2 name. It is the only main-line steam locomotive at Foxfield.

A carriage in the museum has an interesting story. No. 127 was a third-class carriage built in approximately 1880 (by possibly Brown, Marshall & Co.) for North Staffordshire

Railway. With hard wooden seats and only shoulder height partitions, it provided transport for the working class of the time, taking men to most often manual jobs and women to markets. Lighting was provided by oil lamps. In 1907 it was taken out of service. Often decommissioned carriages were removed from their wheels and used as waiting rooms. This one ended up at Waterhouses station, and when the Leek and Manifold line closed (where the station was located) in 1934 the body ended up as a farm building near Ecton. The carriage was rescued and came to Foxfield in 1980. It was eventually restored in 2013 and entered service once again. Upon its return to the track, the first passenger was Mrs Coates, a ninety-five-year-old lady who had used it as a shed on her family farm.

At the beginning of the line in a siding undergoing restoration is a Pacer unit (or Railbus) Class 142 similar to the one that was in use at Midland Railway – Butterley on the day of my visit there. The unit, No. 142055, is being repainted in its original provincial blue. The unit unfortunately suffered vandalism in June 2022, which has put back its restoration schedule. It is owned by the 142055 group and will hopefully soon be up and running carrying passengers.

Further along the line is a level crossing at Cresswellford, where Caverswall Road crosses the railway line linking the villages of Caverswall and Dilhorne. The area has a slightly

Locomotive No. 3 *Marston Thompson Evershed*.

macabre claim to fame, as it is the site of several air crashes in the 1940s and 1950s. On 31 January 1944 an Armstrong Whitworth 'Whitley' bomber from RAF Tilstock crashed into the hill above Foxfield Colliery, nearly knocking the top off the colliery chimney. Men from the colliery rushed to the scene to help, but only the rear gunner, Tom Weightman, survived. He was shot down in 1945 and died in 2007.

The line terminates at Dilhorne Park station, a small halt (used as Hanbury Halt) in the costume drama *Return to Cranford*. The train returns back down the line to Caverswall Road station. It was near Dilhorne Park station that a line branched off up to sidings known as 'Bank Top' and was even steeper than the notorious 'Foxfield Bank'.

A short drive away a colliery tour is available taking you around the buildings of the original Foxfield Colliery. Separate from the railway, it cannot be reached by the railway and is offered on selected dates so it is best to check availability before travelling to the site. The colliery was worked between 1880 and 1965. The main working shaft went to a depth of 752 feet in 1887. The second, a ventilation shaft, is 711 feet deep. The colliery reached peak production in the immediate post-war years employing 390 people in 1947.

A gift shop, café and museum are among the facilitates at Caverswall Road station.

Footplate of Locomotive No. 3.

An open carriage in use at Foxfield, our transport for the journey.

Locomotive *North Stafford No. 2.*

The restored carriage 127.

A footplate showing steam-engine controls.

A 'Signalman's Chair'.

Florence No. 2.

A Class 142 undergoing restoration at Foxfield.

Chapter 10

The Train without a Track: Outside Crewe Heritage Centre

A while back I travelled on a main-line train, not to get to a destination but to travel on a particular locomotive and carriage set, a bit like the late 1980s when as previously explained we chased the last Class 45s across the Midlands.

The train in question was a Pendolino 390 developed by Fiat Ferroviaria. The reason for my interest in this train is the fact the patents for the APT (Advanced Passenger Train), developed by British Rail in the 1970s, were sold to Fiat, who perfected the tilting mechanism and in true entrepreneurial style, sold it back to Britain in the early 2000s as the 390. Virgin bought the locomotives and ran them, doing a publicity photoshoot in 2002 with it at Crewe next to a surviving APT set. Acknowledging its pedigree, even the colour schemes were similar. The short 'hop' we did was from Lichfield Trent Valley to Stafford. According to a GPS app we hit a 125 mph and we experienced the tilt on a high-speed bend. Even after twenty years in service they still look ultra-modern with their sleek lines and folding steps. They are now a blue/green colour since the Avanti take over from Virgin in 2019. I preferred the silver, red and yellow from the Virgin days of the previously mentioned photoshoot.

I never got to ride the original APT so took the above ride on its spiritual successor. A visit to Crewe was in order to get a look at the original. It is in a siding by the Crewe Heritage Centre. The heritage centre was shut when I visited, however I knew the train was visible from a public area.

In a promotional leaflet I tracked down for the original APT the seat allocation was given as:

First Class (Smoking) – 25
First Class (Non Smoking) – 47
Second Class (Smoking) – 52
Second Class (Non Smoking) – 100

Dining Car – 43
Wheelchair – 1

Total seats 268

A total of seventy-seven smoking seats (if the dining car was totally non-smoking) therefore meant around 30 per cent of seats were allocated to smokers. Another question is where the smoking and non-smoking carriages separate. Smoking/non-smoking seating in cinemas of the era were just left and right allocated and the smoke just drifted across. Just one seat space was given to a wheelchair user.

The leaflet also claims a top speed of 125 mph, exactly the same speed as the trip I made on the Pendolino, although the Pendolino seems to hit the speed more consistently. However, on 20 December 1979 British Rail claimed a new speed record of 160 mph. The Intercity 125, developed as a stopgap in 1975 until the APT was ready to enter regular service, also ran at 125 mph (hence the name). It seems high-speed rail travel hasn't really advanced much. The APT record stood until testing for Eurostar started. On what was then known as 'The Channel Tunnel Rail Link' (now known as HS1) the Eurostar hit 203 mph. We have travelled on Eurostar back in 2001 when I took my wife to Paris for her birthday (quite the romantic) and I do remember it having quite a turn of speed, but mainly in France. Around Kent it didn't seem much faster than normal rail travel, although I would have loved a GPS app to check its speed.

I believe the current speed of 125 mph on most lines is governed by signalling limitations on regular routes rather than train design.

The speeds of the APT were monitored on board by British Rail engineers; dot-matrix printers screeched noisily and produced metres of green and white striped paper with the speed on it. A reliable (but anonymous) source tells the story that the numbers were in kph not mph and reversed so hopefully the driver wasn't aware it was his speed being monitored and thought the numbers were more scientific (or more gibberish) than his speed as the numbers bore little resemblance to what he was seeing on his own instruments.

The main feature of the APT was its active tilt feature, which used hydraulic rams that kept the carriages centred over the bogies (carriage wheel sets) so the train could travel up to 40 per cent faster around bends than regular. Some passengers complained of sea sickness though.

The units used on the passenger service were in fact APT-P, which stood for Advanced Passenger Train – Prototype. Only three sets were built along with an earlier set APT-E, which stood for Advanced Passenger Train – Experimental. The APT-P was electric, while the original APT-E was gas turbine.

Perhaps the APT intrigues me so much because when I was growing up, from around seven or eight to my mid-teens, it always seemed to be on the news, not always for the right reasons. I remember it well, because I grew up near to where it was built in Derby so it was always on BBC's *Midlands Today* or ATV's *Today* programme and later *Central News*. I still remember the news bulletin from December 1981 featuring the inaugural running of the APT passenger service. The train was full of journalists from the papers and film crews, and the crew wore smart shiny uniforms that seemed to have an awful lot of 'Crimplene' in them.

A railway forum thread about the APT recalled memories of people who had been fortunate, or unfortunate due to the frequent breakdowns – reliability or rather lack of it seemed to be a recurring theme.

In Britain we love the underdog. The APT was in service for only a few short years but we remember it as a piece of exceptional British workmanship let down by the available technology of the time – a heroic failure as it were. What happened to the APT project? It was quietly taken out of service in 1986 after the government withdrew funding. In the intervening years between initial testing and 1986, the Intercity 125 had entered service on a large scale and formed the backbone of British Rail's high-speed travel for many years.

One of the remaining APT sets is a stationary exhibit at Crewe Heritage Centre and can be viewed (and at certain times can be boarded). It is in a siding but has not turned a wheel for years. The rest of the museum is excellent as well. You can, though, ride the next best thing, the Pendolino, as I did; I was sat among commuters, shoppers and students, who thought they were just on any old train. Could they see the grin on my face and were they wondering why I looked so excited? In my head I was in 1981 riding an APT.

An APT locomotive.

An APT locomotive.

The complicated tilting bogey of the APT.

CHAPTER 11

A Track without a Train: National Memorial Arboretum, Near Alrewas, Staffordshire

Predominantly I've visited and written about railways that are running trains. My last chapter is the most poignant piece of track in the United Kingdom, and it will never see a train run on it again. A length of track with just twenty-five sleepers. Twenty-five, remember that number, it's important.

The track in question is at the National Memorial Arboretum (NMA) in Alrewas in Staffordshire but its original location was thousands of miles away as part of what was known as 'The Death Railway'. I walked along the track in early spring sunshine, a far cry from the blistering heat that the original workers toiled in when it was being built in 1942/43. A grim statistic is that it is believed one prisoner of war or slave labourer died for every sleeper laid. A 100,000 lives were lost – 12,000 Allied prisoners of war and around another 90,000. Each of the twenty-five sleepers relocated in Staffordshire represent one death in the Far East eighty years ago.

I first learnt about the Siam–Burma Railway from the novel *A Town Like Alice* by Nevil Shute after reading it in English at school. While my friends found it a boring love story, I devoured every word, then went on to other books by the same author in my spare time. In *A Town Like Alice*, while a work of fiction, all the information about the war is based on fact. (If you read the book and want to follow it with others by the same author, I would recommend *On the Beach*, one of the first post-apocalyptic novels.)

The arboretum opened in 2001 and the track was brought over soon after and dedicated as a memorial in 2002. Thirty people helped unload the track and manoeuvre it into position, including some veterans from the Far East campaign. I can imagine it was very emotional for them remembering fallen comrades with whom they had toiled alongside many decades before.

Home for the prisoners of war during the war was often a construction camp with open-sided barracks made of bamboo, 60 metres long holding 200 prisoners. They were placed at intervals of around 5 to 10 miles along the route. Food was scarce, which meant malnutrition and starvation for many of the workers. Serious illness was met with disinterest; you went to the camp hospital and either got better or died. Little was offered in terms of medication or sympathy by the enemy.

There were 600 bridges needed over the 258-mile route. Eight of them were long-span bridges like the most famous of them, the bridge on the River Kwai, which was the subject of a fictional novel and film by Pierre Boulle. In real life the bridge wasn't destroyed by the British and a distraught Alec Guinness. It was a temporary wooden bridge but before the end of the war it was replaced by a concrete and steel structure which was the target for a new weapon, the Azon Bomb, an early guided bomb that could be steered left or right as it fell. The mission was aborted due to bad weather and the bridge still stands today.

After the railway, many prisoners were held at Changi prison. A lychgate was built there by Royal Engineers as an entrance to the burial site at the camp. Following the closure of the camp in 1945, it was put into storage. It was erected in 1971 at the barracks in Bassingboune, Cambridge. It stood there until 2003 when it was transported to the NMA. Once restored, it became the entrance to the Far East Memorial area.

The piece is an original piece of track recovered from the Burma Railway. It was manufactured, however, in Middlesborough and exported in the 1930s before hostilities began. There is another length of track as a memorial to the Sumatra Railway. This piece, however, is ex-British Rail stock and was sourced for the memorial in the UK.

Within the Far East area of the arboretum is a hidden memorial, a piece of land with the right angle formed by the two tracks. The sunken area is the size of a tennis court and is a memorial to the Battle of Kohima fought in North East India in 1944. Part of the battle was

Length of Burma–Siam Railway.

fought over an actual tennis court in the grounds of the Deputy Commissioner's residence during a siege, the skirmish becoming known as the Battle of the Tennis Court.

A permanent exhibition hall is dedicated to the Far East Campaign and is the largest on the site after the main building. Stories of casualties and survivors are told with images depicting the harsh conditions suffered by the prisoners of war who built both the railways represented by the lengths of track.

Elsewhere at the National Memorial Arboretum is a Railway Industry Memorial dedicated to railway workers who lost their lives in both wartime and peacetime working on the railway. Twice a year a wreath is laid, one on Armistice Day and the other on 22 May, the anniversary of the dedication in 2012. This memorial is within sight of the Derby to Birmingham line, the line that serves my hometown.

A week would be needed to see every memorial at the NMA, and while most aren't railway related, the history and stories are fascinating and thought provoking. The Shot at Dawn memorial and the Armed Forces Memorial (known commonly as 'The Wall') are truly heartbreaking.

Lychgate from Changi prison, built by Allied prisoners of war during the Second World War.

Far East Prisoners of War Memorial building at NMA.

'Hidden' memorial to the Battle of the Tennis Court.

Epilogue

I won't end this book on railways with the sadness of death and warfare. Heritage railways are memorials in themselves, a tribute to British industry and ingenuity. Volunteers give up their free time to run trains, both steam and diesel.

Visit them when you can and smile at the sight and smells of the engines, with the staff often dressed in period costumes – eat and drink in a bygone age. If you have smaller children or grandchildren, take them on a Santa Special. A mince pie, a glass of sherry (for you obviously) and a present handed over by the man himself on a steam train, what could be a better Christmas treat?

Don't just visit the railways in this book. Seek out different ones while on holiday and look for others on your doorstep and ride the rickety narrow gauges. Each one will have different engines and a unique story to tell, and if you find a Class 45 running let me know.